T0225734

An Introduction to Smooth Manifolds

Manjusha Majumdar · Arindam Bhattacharyya

An Introduction to Smooth Manifolds

 Springer

Manjusha Majumdar
Department of Pure Mathematics
University of Calcutta
Kolkata, West Bengal, India

Arindam Bhattacharyya
Department of Mathematics
Jadavpur University
Kolkata, West Bengal, India

ISBN 978-981-99-0567-6   ISBN 978-981-99-0565-2   (eBook)
https://doi.org/10.1007/978-981-99-0565-2

Mathematics Subject Classification: 58-01, 58AXX, 58A05, 58A10, 58A15, 58C15, 58C25

This Springer imprint is published by the registered company Springer Nature Singapore Pte Ltd.
The registered company address is: 152 Beach Road, #21-01/04 Gateway East, Singapore 189721,
Singapore

Dedicated to
Gahana Majumdar and Bipasa
Bhattacharyya

Preface

This book is an outcome of lectures delivered by one of the authors to the postgraduate students at the Department of Pure Mathematics, University of Calcutta, India, for several years. The need of the students has motivated the authors to write a textbook. The only prerequisites are good working knowledge of point-set topology and linear algebra.

It is said that mathematics can be learnt by solving problems, not only by reading it. To serve this purpose, this book contains a sufficient number of examples and exercises after each section of every chapter. Some exercises are routine ones for the general understanding of each section. We have given hints about difficult exercises. Answers to all exercises are given at the end of each section. We hope that this approach will help the readers for getting this beautiful subject accessible.

We do not believe that there can be any complete book on the topic of manifold. We are sure our book is far from completion as such. However, we are equally sure that our book has some exceptional merits, and students will be benefitted if they go through the whole book with all exercises.

Chapter 1 is the study of calculus on $\mathbb{R}^n$. We have started the first section on smooth functions. The concept of the diffeomorphic function is as important as diffeomorphic manifold. We have given a few exercises on that. Tangent vector is one of the powerful concepts of studying geometry. It has been defined with respect to a curve in $\mathbb{R}^n$ in the second section.

The germ of a function has been defined in the third section. The last two sections are on inverse function theorem and implicit function theorem with examples and exercises.

Chapter 2 is the study of manifold. We have defined topological manifold and then smooth manifold in Sect. 2.3. Many exercises have been given for a better understanding of the concept of atlas of smooth manifold. Germs on topological manifold have been explained in Sect. 2.2. Stereographic projections and orientable surfaces are two attractive concepts, which have been explained separately in Sect. 2.4 and 2.5, respectively. Product manifold has been explained separately in Sect. 2.6. Smooth functions on smooth manifold have been explained with solved problems in Sect. 2.7. In this section, we have included diffeomorphic smooth manifold. Tangent vector has been introduced with respect to a differentiable curve in Sect. 2.8. The next section is the study of inverse function theorem for smooth manifold. Section 2.10 deals with vector field and its geometrical interpretation has been explained in the next section. Section 2.12 deals with push-forward vector fields which lead to the concept of submanifold and hence to critical points and regular points of the manifold. We have discussed Submanifolds separately in the next section. Push-forward vector fields give rise to another kind of vector field which has been explained with solved problems in Sect. 2.14. Finally, the last section of this chapter gives the algebraic interpretation of the vector field. It is also termed as flow while studying dynamical system which is an interesting topic of mathematics.

Chapter 3 is the study of differential forms. Differential forms have wide applications in Lie group, differential topology, differentials and their multiple integrals over a differentiable manifold. However, in this book, we have mainly considered the first role played by differential forms on manifolds.

The first section of this chapter is on 1-form, which is also called cotangent space. It can be thought of as dual vector space of tangent space of the manifold. Thus, tangent space and cotangent space can be thought of as "siamese twins" at every point of the manifold. Members of cotangent space are also called co-vectors. We have given the formal definition of r-form ($r > 1$) in the next section. Differential r-forms are tensor fields of type (0, r) which are skew-symmetric. They have wide applications in thermodynamics. We have also studied exterior product or wedge product in this section, which is nothing but the generalization of the concept of cross-product between two vectors in 3-dimension. This beautiful concept was introduced by R. G. Grassmann, which nowadays also called "Grassmann algebra" and the reason for this name "algebra" has been explained in this section for those students who are of inquisitive nature. Exterior derivative to a manifold is the same as that of "curl" to R3. All the classical concepts, namely gradient and divergence, can be expressed in terms of this concept. We have given the proof of existence and uniqueness of such operations. Finally, pull-back differential form has been studied in the last section. This pull-back operation and exterior differentiation commute each other which has been explained by a theorem, followed by many exercises.

The last chapter is on Lie group. The Lie group structure allows us to discuss continuity and differentiability in a group structure. It was introduced by Norwegian mathematician S. Lie in the late nineteenth century. Lie groups play an important role in modern geometry. They are the fundamental building blocks for Gauge theories.

The first section is the study of Lie group and the two C^∞ transformations on it. The behaviour of a Lie group is determined by its behaviour in the neighbourhood of its identity element, and hence a famous theorem has been studied in the next section. Due to the two translations, two types of invariant vector fields occur in this group. Naturally, two types of invariant differential forms are also there. Well-known theorems and results have been studied in the next two sections.

For the unique structure of a Lie group, one should have a natural quest for studying group homomorphism and algebra homomorphism on it. The unique feature is the study of one-parameter group of transformations induced by the invariant vector field of a Lie group. Section 4.6 is the study of the action of a Lie group on a manifold.

We have gone through many books and articles during the preparation of this book, which have been listed in Bibliography.

We wish to express our sincere acknowledgement to our respected teachers, colleagues and friends for their valued suggestions.

One of the authors of this book would like to express her deep gratitude to her deceased teacher, Prof. M. C. Chaki, University of Calcutta. His wonderful lectures on this beautiful subject always generated great excitement and enthusiasm for the students. Moreover, his affection, guidance and blessings towards her are memorable forever.

We invite suggestions and comments from our readers for incorporation, if any, in future editions.

We wish to thank Dr. S. Kundu, Assistant Professor, Loreto College, Kolkata, India, for his excellent work on typing the entire manuscript. He has offered some valuable suggestions also.

We also wish to thank the budding mathematician Mr. Sumanjit Sarkar, Research Scholar, under Prof. A. Bhattacharyya, for his tireless effort in re-checking the whole manuscript.

Finally, we acknowledge Mr. Shamim Ahmed, Springer, New Delhi, who took all necessary care and perseverance for the preparation of this book.

Both authors also wish to convey their regards to the respected reviewers for the enrichment of this book.

Kolkata, India   Manjusha Majumdar
2022   Arindam Bhattacharyya

Contents

About the Authors

Manjusha Majumdar, Ph.D., is former Professor of Pure Mathematics at the University of Calcutta, India. Her main interest lies in smooth manifolds. She has published a number of research papers in several international journals of repute. Under her supervision, 10 students have already been awarded their Ph.D. degrees. She is a member of many national and international mathematical societies and serves on the editorial board of several journals of repute. She has visited several institutions in India and abroad on invitation. She is the Principal Investigator of the prestigious e-PG Pathshala in Mathematics and MOOC in Mathematics, recommended by the Ministry of Human Resource Development, Government of India. She co-authored a book entitled *Differential Geometry*.

Arindam Bhattacharyya, Ph.D., is Professor, Department of Mathematics, Jadavpur University, Kolkata, India. His main interest lies in smooth manifolds, geometric flows and computational geometry. He has published a number of research papers in several international journals of repute. Under his supervision, 19 research scholars have already been awarded their Ph.D. degrees and 8 are pursuing their research. He is a member of many national and international mathematical societies and serves on the editorial board of several journals of repute. He has visited several institutions in India and abroad on invitation. He has organized several international conferences and research schools in collaboration with internationally reputed mathematical organizations. He is the Course Coordinator of Differential Geometry of the prestigious e-PG Pathshala in Mathematics and MOOC in Mathematics, recommended by the Ministry of Human Resource Development, Government of India, and of Netaji Subhas Open University, India. He co-authored a book entitled *Differential Geometry* with Prof. Manjusha Majumdar.

List of Figures

Chapter 1
Calculus on $\mathbb{R}^n$

1.1 Smooth Functions

Let $\mathbb{R}$ denote the set of real numbers. For an integer $n > 0$, let $\mathbb{R}^n$ denote the set of all ordered n-tuples $(x^1, x^2, \ldots, x^n)$ of real numbers. Individual n-tuple will be denoted at times by a single letter, e.g. $x = (x^1, x^2, \ldots, x^n)$, $y = (y^1, y^2, \ldots, y^n)$ and so on.

A real-valued function $f : U \subset \mathbb{R}^n \to \mathbb{R}$, U being an open set, is said to be of class C^k if the following conditions hold:

(i) all its partial derivatives of the order less than or equal to k exist, and
(ii) are continuous functions on U.

By class C^∞, we mean that all orders of partial derivatives of f exist and are continuous at every point of U. A function of class C^∞ is also called a smooth function. Actually, "Smoothness" is a synonym for C^∞. By class C^0, we mean that f is merely continuous from U to $\mathbb{R}$. By class C^ω on U, we mean that f is real analytic on U. A C^ω function is C^∞ function but the converse is not true.

Example 1.1 Let $f : \mathbb{R} \to \mathbb{R}$ be defined by $f(x) = x^{\frac{1}{3}}$. Then

$$f'(x) = \begin{cases} \frac{1}{3}x^{-\frac{2}{3}}, & x \neq 0; \\ 0, & x = 0. \end{cases}$$

Hence, f is C^0 but not C^1.

Example 1.2 The polynomial, sine, cosine and exponential functions on the real line are all C^∞, which are also analytic, i.e. C^ω.

© The Author(s), under exclusive license to Springer Nature Singapore Pte Ltd. 2023
M. Majumdar and A. Bhattacharyya, *An Introduction to Smooth Manifolds*,
https://doi.org/10.1007/978-981-99-0565-2_1

Problem 1.1 *Let $f : \mathbb{R} \to \mathbb{R}$ be defined by*

$$f(x) = \begin{cases} e^{-\frac{1}{x^2}}, & x \neq 0; \\ 0, & x = 0. \end{cases}$$

Show that f is a function of class C^∞.

Solution: Note that

$$f'(0) = \lim_{h \to 0} \frac{f(0+h) - f(0)}{h} = \lim_{h \to 0} \frac{e^{-\frac{1}{h^2}}}{h}.$$

Put $h = \frac{1}{u}$, then

$$f'(0) = \lim_{u \to \infty} ue^{-u^2} \left(\frac{\infty}{\infty} \right).$$

Using L'Hospital's rule, we find

$$f'(0) = \lim_{u \to \infty} \frac{1}{2ue^{u^2}} = \lim_{u \to \infty} \frac{e^{-u^2}}{2u} = 0.$$

Again

$$f'(x) = \frac{2}{x^3} e^{-\frac{1}{x^2}}.$$

Therefore

$$f''(0) = \lim_{h \to 0} \frac{\frac{2}{h^3} e^{-\frac{1}{h^2}}}{h} = \lim_{u \to \infty} \frac{2e^{-u^2}}{\frac{1}{u^4}}, \quad \text{taking } u = \frac{1}{h}.$$

Using L'Hospital's Rule successively, we will find

$$f''(0) = 2 \lim_{u \to \infty} \frac{24u}{4ue^{u^2} + 8ue^{u^2} + 8u^3 e^{u^2}}.$$

Finally, we will find

$$f''(0) = 0.$$

Proceeding in the same manner, we will find

$$f^n(0) = 0, \quad n = 1, 2, 3, \ldots \ldots$$

Hence, we claim that f is a function of class C^∞.

Example 1.3 The function defined in Example 1.1 above, does not have power series expansion at $x = 0$. Hence it is not a C^ω function.

Problem 1.2 *Consider the functions*

$$f_1(x) = \begin{cases} 0, & x \le 1; \\ e^{-\frac{1}{(x-1)^2}}, & x > 1. \end{cases}$$

$$f_2(x) = 0, \ for \ -\infty < x < \infty.$$

Prove that f_1, f_2 are differentiable on $\mathbb{R}$.

Solution: For $x \le 1$, $f_1(x) = 0$, so $f_1'(1) = 0$. For $x > 1$, $f_1'(x) = 2(x-1)e^{-\frac{1}{x^2}}$.
Now

$$\begin{aligned} Rf_1'(1) &= \lim_{h \to 0} \frac{f(1+h) - f(1)}{h} = \lim_{h \to 0} \frac{e^{-\frac{1}{h^2}}}{h} \\ &= \lim_{u \to \infty} \frac{u^{\frac{1}{2}}}{e^u}, \quad u = \frac{1}{h^2} \\ &= 0. \end{aligned}$$

$$Lf_1'(1) = \lim_{h \to 0} \frac{f(1-h) - f(1)}{h} = \lim_{h \to 0} \frac{e^{-\frac{1}{h^2}}}{h} = 0.$$

Thus, $f_1'(1)$ exists and continuous. So f_1 is differentiable on $\mathbb{R}$.

As $f_2(x)$ is a constant function, it has finite derivative everywhere and so f_2 is differentiable on $\mathbb{R}$.

Problem 1.3 *Let $f : \mathbb{R} \to \mathbb{R}$ be defined by $f(x) = x^{\frac{4}{3}}$. Show that the function g defined by*

$$g(x) = \int_0^x f(t)dt = \int_0^x t^{\frac{4}{3}}dt = \frac{3}{7}x^{\frac{7}{3}},$$

is C^2 but not C^3.

Solution: Now $g'(x) = f(x) = x^{\frac{4}{3}}$. Thus, $g(x)$ is C^1. Again $g''(x) = \frac{4}{3}x^{\frac{1}{3}}$, so $g(x)$ is C^2 but not C^3 at $x = 0$.

Exercises

Exercise 1.1 *Let $f : \mathbb{R} \to \mathbb{R}$ be defined by*

$$f(x) = \begin{cases} e^{-\frac{1}{x}}, & x > 0; \\ 0, & x \le 0. \end{cases}$$

Show that f is a function of class C^∞.

Exercise 1.2 *Define $f : \mathbb{R} \to \mathbb{R}$ by $f(x) = x^3$. Is f^{-1} of class C^∞?*

Exercise 1.3 *Let* $f : \mathbb{R} \to \mathbb{R}$ *be defined by*

$$f(x) = \begin{cases} x + x^2 \cos \frac{1}{x}, & x \neq 0; \\ 0, & x = 0. \end{cases}$$

Show that

(i) f *is continuous.*
(ii) f *is differentiable at all points.*
(iii) *the derivative is discontinuous at* $x = 0$.

Exercise 1.4 *Let* $f : \mathbb{R} \to \mathbb{R}$ *be defined by*

$$f(x) = \begin{cases} x^2 \sin \frac{1}{x}, & x \neq 0; \\ 0, & x = 0. \end{cases}$$

Show that f *is differentiable at* $x = 0$ *but* f' *is not continuous at* $x = 0$.

Answers

1.2 No.

For $i = 1, 2, \ldots, n$, let $u^i : \mathbb{R}^n \to \mathbb{R}$ be the natural co-ordinate functions i.e.

$$u^i(p) = p^i, \text{ where } p = (p^1, p^2, \ldots, p^n) \in \mathbb{R}^n. \tag{1.1}$$

Such u^i's are continuous functions from $\mathbb{R}^n$ to $\mathbb{R}$ and we call this n-tuple of functions $u^1, u^2, \ldots, u^n$; the standard co-ordinate system of $\mathbb{R}^n$. If $f : U \subset \mathbb{R}^n \to \mathbb{R}^n$ is a mapping, then f is determined by its co-ordinate functions $f^1, f^2, \ldots, f^n$ where

$$f^i = u^i \circ f, \ i = 1, 2, 3, \ldots, n \tag{1.2}$$

and each f^i is a real-valued function. Thus

$$f(p) = (f^1(p), f^2(p), \ldots, f^n(p)) \ \forall \ p \in \mathbb{R}^n. \tag{1.3}$$

The map f is of class C^∞ if each component function f^i, $i = 1, 2, 3, \ldots, n$ is also so.

If $f : U \subset \mathbb{R}^n \to \mathbb{R}^m$ is a mapping such that

$$f(x^1, x^2, \ldots, x^n) = (f^1(x^1, x^2, \ldots, x^n), f^2(x^1, x^2, \ldots, x^n), \ldots, f^m(x^1, x^2, \ldots, x^n)),$$

we define the Jacobian matrix of f at $(x^1, x^2, \ldots, x^n)$, denoted by J, as

$$J = \begin{pmatrix} \frac{\partial f^1}{\partial x^1} & \frac{\partial f^1}{\partial x^2} & \cdots & \frac{\partial f^1}{\partial x^n} \\ \frac{\partial f^2}{\partial x^1} & \frac{\partial f^2}{\partial x^2} & \cdots & \frac{\partial f^2}{\partial x^n} \\ \cdots & \cdots & \cdots & \cdots \\ \frac{\partial f^m}{\partial x^1} & \frac{\partial f^m}{\partial x^2} & \cdots & \frac{\partial f^m}{\partial x^n} \end{pmatrix},$$

provided each of its first-order partial derivatives exist on U.

If $m = 1$, the matrix $\left(\dfrac{\partial f}{\partial x^1} \dfrac{\partial f}{\partial x^2} \cdots \dfrac{\partial f}{\partial x^n} \right)$ is called the gradient of f, denoted by 'grad f' or ∇f.

The function f is said to be continuously differentiable on U, if each f^i has first-order continuous partial derivatives on U. If f is continuously differentiable on U and the Jacobian is non-null, then f is one-to-one in U.

A mapping $f : U \subset \mathbb{R}^n \to V \subset \mathbb{R}^n$, U, V being open sets in $\mathbb{R}^n$, is said to be homeomorphism if

(i) f is bijective and
(ii) f, f^{-1} are continuous.

Problem 1.4 *Let $f : \mathbb{R} \to \mathbb{R}$ be such that $f(x) = 5x + 3$. Show that f is a homeomorphism on $\mathbb{R}$.*

Solution: Here

$$f(x) - f(y) = 5(x - y)$$
$$\Rightarrow f(x) = f(y) \text{ if and only if } x = y.$$

Thus, f is one-to-one. Further to examine whether f is onto, we are to examine if there exists a pre-image x of y under f such that $f(x) = y$ holds. So y has a pre-image $\frac{y-3}{5}$ in the domain set $\mathbb{R}$. Since y is arbitrary, each element in the domain set $\mathbb{R}$ has a pre-image under f. Thus, f is onto. Hence, f is bijective so $f^{-1} : \mathbb{R} \to \mathbb{R}$ exists, which is defined by

$$f^{-1}(x) = \frac{x - 3}{5}.$$

Here f^{-1} is continuous. Also, f is continuous. Consequently, f is a homeomorphism.

Problem 1.5 *Let $f : \mathbb{R} \to \mathbb{R}$ be such that $f(x) = x^3$. Test*

(i) *whether f is of class C^∞ or not.*
(ii) *whether f is a homeomorphism or not.*

Solution:

(i) Here

$$f'(0) = \lim_{h \to 0} h^2 = 0$$
$$f'(x) = 3x^2.$$

Thus f has a finite derivative and f is of class C^∞.

(ii) Here

$$f(x) - f(y) = x^3 - y^3$$
$$\Rightarrow f(x) = f(y) \text{ if and only if } x = y.$$

Thus, f is one-to-one. Further to examine whether f is onto, we are to examine if there exists a pre-image x of y under f such that $f(x) = y$ holds. So y has a pre-image $y^{\frac{1}{3}}$ in the domain set $\mathbb{R}$. Since y is arbitrary, each element in the domain set $\mathbb{R}$ has a pre-image under f. Thus, f is onto. Hence, f is bijective so $f^{-1} : \mathbb{R} \to \mathbb{R}$ exists, which is defined by

$$f^{-1}(x) = x^{\frac{1}{3}}.$$

Here f^{-1} is continuous. Also, f is continuous. Consequently, f is a homeomorphism.

Exercises

Exercise 1.5 *Let $f : \mathbb{R} \to \mathbb{R}$ be such that $f(x) = x^2$. Is f a homeomorphism on $\mathbb{R}$?*

Exercise 1.6 *The function $f : (-1, 1) \to \mathbb{R}$ is defined by $f(x) = \dfrac{x}{1 - x^2}$. Is f a homeomorphism on $(-1, 1)$?*

Exercise 1.7 *Let $f : \left(-\dfrac{\pi}{2}, \dfrac{\pi}{2}\right) \to \mathbb{R}$ be such that $f(x) = \tan x$. Show that f is a homeomorphism.*

Exercise 1.8 *Let $f : \mathbb{R}^2 \to \mathbb{R}^2$ be such that $\phi(u, v) = (ve^u, u)$. Is ϕ a homeomorphism on $\mathbb{R}^2$?*

Exercise 1.9 *Consider the mapping $\phi : \mathbb{R}^3 \to \mathbb{R}^3$ given by*

$$\phi(x^1, x^2, x^3) = (x^1 + 1, x^2 + 2, x^3 + 3).$$

Show that ϕ is a homeomorphism.

Answers

1.5 No.   **1.6** Yes.   **1.8** Yes.

Remark 1.1 A homeomorphism between open subsets of $\mathbb{R}^n$ and $\mathbb{R}^m$, $n \neq m$ is not possible. For details, refer to any book on topology.

A mapping $f : U \subset \mathbb{R}^n \to V \subset \mathbb{R}^n$, U, V being open sets in $\mathbb{R}^n$, is said to be diffeomorphism if

  (i) f is a homeomorphism of U onto V and
  (ii) f, f^{-1} are of class C^∞.

Problem 1.6 *Let $f : \mathbb{R} \to \mathbb{R}$ be such that $f(x) = x^3$. Test whether f is a diffeomorphism or not.*

Solution: We have shown in Example 1.5 that f is a homeomorphism and f is of class C^∞. Now

$$f^{-1}(x) = x^{\frac{1}{3}}.$$

Such f^{-1} is not C^1, as $(f^{-1}(x))' = \frac{1}{3}x^{-\frac{2}{3}}$, which is not defined at $x = 0$. Thus f^{-1} is not of class C^∞. Consequently, f is not a diffeomorphism.

Problem 1.7 *Let $f : \mathbb{R}^2 \to \mathbb{R}^2$ be such that $f(u, v) = (ue^v + v, ue^v - v)$. Show that f is a diffeomorphism.*

Solution: Note that

$$|J| = \begin{vmatrix} \frac{\partial f^1}{\partial u} & \frac{\partial f^1}{\partial v} \\ \frac{\partial f^2}{\partial u} & \frac{\partial f^2}{\partial v} \end{vmatrix} = \begin{vmatrix} e^v & ue^v + 1 \\ e^v & ue^v - 1 \end{vmatrix} = -2e^v \neq 0, \ \forall \, v.$$

Hence, f is invertible. If

$$(\xi, \eta) = f(x^1, x^2), \quad \text{then} \ \ \xi = x^1 e^{x^2} + x^2, \ \eta = x^1 e^{x^2} - x^2.$$

Thus $\xi + \eta = 2x^1 e^{x^2}$ and $\xi - \eta = 2x^2$. Consequently,

$$x^2 = \frac{1}{2}(\xi - \eta) \ \text{ and } \ x^1 = \frac{1}{2}(\xi + \eta)e^{\frac{1}{2}(\eta - \xi)}.$$

We now define

$$f^{-1}(\xi, \eta) = \left(\frac{1}{2}(\xi + \eta)e^{\frac{\eta - \xi}{2}}, \frac{1}{2}(\xi - \eta) \right).$$

Also, f, f^{-1} is continuous, hence homeomorphism. Also, both f, f^{-1} are of class C^∞. So f is diffeomorphism.

Problem 1.8 *Let $f : \mathbb{R}^3 \to \mathbb{R}^3$ be such that $f(u, v, \omega) = (ue^\omega + ve^\omega, ue^\omega - ve^\omega, \omega)$. Test whether f is a diffeomorphism or not at $(1, 1, 0)$.*

Solution: Note that

$$|J| = \begin{vmatrix} e^\omega & e^\omega & ve^\omega + ue^\omega \\ e^\omega & -e^\omega & -ve^\omega + ue^\omega \\ 0 & 0 & 1 \end{vmatrix} = -e^{2\omega} \neq 0 \text{ at } (1, 1, 0).$$

Thus, f is invertible at $(1, 1, 0)$. Again, if $(\xi, \eta, \theta) = f(u, v, \omega)$, then

$$\xi + \eta = 2ue^\omega \Rightarrow u = \frac{1}{2}(\xi + \eta)e^{-\omega}$$

$$\xi - \eta = 2ve^\omega \Rightarrow v = \frac{1}{2}(\xi - \eta)e^{-\omega}$$

$$\theta = \omega.$$

We now define

$$f^{-1}(\xi, \eta, \theta) = \left(\frac{\xi + \eta}{2}e^{-\theta}, \frac{\xi - \eta}{2}e^{-\theta}, \theta \right).$$

Also f, f^{-1} is continuous, hence homeomorphism at $(1, 1, 0)$. Also, both are of class C^∞. So f is diffeomorphism at $(1, 1, 0)$.

Exercises

Exercise 1.10 Let $f : \left(-\frac{\pi}{2}, \frac{\pi}{2}\right) \to \mathbb{R}$ be defined by $f(x) = \tan x$. Show that f is a diffeomorphism.

Exercise 1.11 Let $\phi : \mathbb{R}^2 \to \mathbb{R}^2$ be defined by $\phi(u, v) = (ve^u, u)$. Determine whether ϕ is a diffeomorphism or not.

Exercise 1.12 Consider the map $f : \mathbb{R}^2 \to \mathbb{R}^2/\{0, 0\}$ defined by $f(x, y) = (e^x \sin y, e^x \cos y)$.

(i) Prove that Jacobian determinant of f does not vanish at any point of $\mathbb{R}^2$.
(ii) Is f a diffeomorphism?

Exercise 1.13 Let $\phi : \mathbb{R}^3 \to \mathbb{R}^3$ be the map defined by

$$x^1 = e^{2x^2} + e^{2x^3}$$
$$x^2 = e^{2x^1} - e^{2x^3}$$
$$x^3 = x^1 - x^2.$$

Show that ϕ is a diffeomorphism.

Exercise 1.14 Consider the C^∞ function $\phi : \mathbb{R}^3 \to \mathbb{R}^3$ defined by

$$\phi(x^1, x^2, x^3) = (x^1 \cos x^3 - x^2 \sin x^3, x^1 \sin x^3 + x^2 \cos x^3, x^3).$$

Prove that $\phi\big|_{S^2}$ is a diffeomorphism from the unit sphere S^2 onto itself.

Exercise 1.15 *Prove that the mapping $\phi : \mathbb{R}^3 \to \mathbb{R}^3$ given by $f(u, v, w) = (u \sin \omega + v \cos \omega, u \cos \omega - v \sin \omega, \omega)$ is a diffeomorphism.*

Answers

1.11 diffeomorphism.   **1.12** (ii) No, f is not one-to-one.

1.2   Tangent Vector

It is known that in $\mathbb{R}^3$, any line, say any curve $\gamma(t)$, through a point $p \in \mathbb{R}^3$, parallel to a non-zero vector v has equation of the form

$$\gamma(t) = p + tv, \quad t \text{ being the parameter.}$$

Thus, we can write

$$\gamma(t) = (p^1 + tv^1, p^2 + tv^2, p^3 + tv^3), \quad p = (p^1, p^2, p^3), \quad v = (v^1, v^2, v^3).$$

Hence, any point, on this curve, has the co-ordinate, say (x^1, x^2, x^3), where

$$x^i \equiv \gamma^i(t) = (p^i + tv^i), \quad i = 1, 2, 3.$$

Thus, in $\mathbb{R}^n$, the curve through $p = (p^1, p^2, \ldots, p^n)$, parallel to the direction of the non-zero vector $v = (v^1, v^2, \ldots, v^n)$ is of the form

$$\begin{cases} \gamma(t) = (p^1 + tv^1, p^2 + tv^2, \ldots, p^n + tv^n), \gamma(0) = (p^1, p^2, \ldots, p^n) \equiv p \in \mathbb{R}^n \\ x^i \equiv \gamma^i(t) = (p^i + tv^i), i = 1, 2, 3, \ldots, n, \ t \text{ being the parameter.} \end{cases}$$

$$(1.4)$$

Let f be a C^∞ function in a neighbourhood of p of $\mathbb{R}^n$. Then the tangent vector at p, in the direction of v, is defined to be the directional derivative, denoted by $D_v f$, as follows:

$$D_v f = \lim_{t \to 0} \frac{f(\gamma(t)) - f(\gamma(0))}{t} = \frac{d}{dt} f(\gamma(t)) \Big|_{t=0}$$

$$= \sum \frac{\partial f(\gamma(t))}{\partial \gamma^i} \Big|_{t=0} \frac{d\gamma^i(t)}{dt} \Big| t = 0$$

$$= \sum \frac{\partial f(p)}{\partial x^i} v^i, \text{ by (1.4).}$$

$$D_v f = \sum v^i \frac{\partial f(p)}{\partial x^i}. \tag{1.5}$$

We also write

$$D_v = \sum v^i \frac{\partial}{\partial x^i}(p). \tag{1.6}$$

Fig. 1.1 Tangent plane

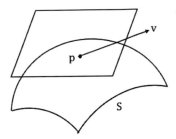

Thus, the directional derivative acts as an operator on functions (Fig. 1.1).

Remark 1.2 Note that in $\mathbb{R}^3$, a vector at p, is a tangent vector, defined by v say, to a surface S in $\mathbb{R}^3$, if it lies on the tangent plane at p.

Thus, the tangent space at p of $\mathbb{R}^n$, denoted by $T_p(\mathbb{R}^n)$, is the collection of all tangent vectors v at p. Such a space is a vector space and hence from the Fundamental Theorem of finite-dimensional Vector Space, any $v \in T_p(\mathbb{R}^n)$ can be expressed uniquely as $v = \sum_{i=1}^{n} v^i e_i$, $\forall v_i \in \mathbb{R}$, where $\{e_1, e_2, \ldots, e_n\}$ is a basis of $T_p(\mathbb{R}^n)$.

Problem 1.9 *Let $v = (2, 3, 0)$ denote a vector in $\mathbb{R}^3$. Find $D_v f$, for a fixed point $p = (-2, \pi, 1)$ where $f : \mathbb{R}^3 \to \mathbb{R}$ is defined by $f = x^1 x^3 \cos x^2$.*

Solution: In this case, $p + tv = (-2 + 2t, \pi + 3t, 1)$. Therefore, $f(p + tv) = 2 \cos 3t - 2t \cos 3t$. Hence,

$$D_v f = \frac{d}{dt} f(p + tv)\Big|_{t=0} = -2.$$

Alternative Method

From (1.5), we obtain $D_v f = \sum_i v^i \frac{\partial f}{\partial x^i}(p)$. In this case

$$D_v f = 2(x^3 \cos x^2)(-2, \pi, 1) + 3(-x^1 x^3 \sin x^2)(-2, \pi, 1)$$
$$= -2.$$

Problem 1.10 *Let $\gamma(t) = \begin{pmatrix} \sin 2t & \cos 2t \\ \cos 2t & -\sin 2t \end{pmatrix} \begin{pmatrix} x \\ y \end{pmatrix}$ be a curve in $\mathbb{R}^3$ with initial point $p \in \mathbb{R}^3$ be such that $\gamma(0) = p$. Find the velocity vector $\gamma'(0)$ at p. Hence compute $D_v f$, where $f : \mathbb{R}^2 \to \mathbb{R}$ is defined by $f = 2x + y^3$.*

Solution: Note that

$$\frac{d}{dt} \gamma(t)\Big|_{t=0} = \frac{d}{dt}(x \sin 2t + y \cos 2t, x \cos 2t - y \sin 2t)\Big|_{t=0} = (2x, -2y).$$

Thus, the velocity vector at p in $\mathbb{R}^2$ is given by

$$2x\frac{\partial}{\partial x} - 2y\frac{\partial}{\partial y}.$$

Hence

$$D_v f = \sum_i v^i \frac{\partial f}{\partial x^i}(p), \quad \text{where } v = (2x, -2y)$$

is given by

$$D_v f = 4x - 6y^3.$$

Alternative Method

In this case

$$p + tv = (x, y) + t(2x, -2y) = ((1 + 2t)x, (1 - 2t)y).$$

Thus

$$f(p + tv) = 2(1 + 2t)x + (1 - 2t)^3 y^3.$$

Consequently

$$D_v f = \frac{d}{dt} f(p + tv)|_{t=0} = 4x - 6y^3.$$

Problem 1.11 Let $p = (1, 1, 0)$ be a point in $\mathbb{R}^3$ and let

$$\gamma_p(t) = (e^t, \cos t, t), \quad t \in \mathbb{R}$$

be a curve with initial point $p \in \mathbb{R}^3$. Find the velocity vector v in $\mathbb{R}^3$ and hence compute $D_v f$, where $f : \mathbb{R}^3 \to \mathbb{R}$ is defined by $f = xz \cos y$.

Solution: Here

$$\frac{d}{dt}\gamma_p(t)\Big|_{t=0} = (e^t, -\sin t, 1)\Big|_{t=0} = (1, 0, 1) = v.$$

Thus the velocity vector at p in $\mathbb{R}^3$ is given by

$$\frac{\partial}{\partial x} + \frac{\partial}{\partial z}.$$

Hence

$$D_v f = \sum_{i=1}^n v^i \frac{\partial f}{\partial x^i}(p) = \cos 1. \tag{1.7}$$

Alternative

Here

$$p + tv = (1, 1, 0) + t(1, 0, 1) = (1 + t, 1, t).$$

Therefore,

$$D_v f = \frac{d}{dt} f(p + tv)|_{t=0} = (\cos 1 + 2t \cos 1)\big|_{t=0} = \cos 1.$$

Exercises

Exercise 1.16 *Let* $v = (2, -3, 4)$ *denote a vector in* $\mathbb{R}^3$. *For a fixed point* $p = (2, 5, 7)$, *compute* $D_v f$ *where*

(i) $f : \mathbb{R}^3 \to \mathbb{R}$ *is defined by* $f = x^3 y$.
(ii) $f : \mathbb{R}^3 \to \mathbb{R}$ *is defined by* $f = z^7$.
(iii) $f : \mathbb{R}^3 \to \mathbb{R}$ *is defined by* $f = e^x \cos z$.

Exercise 1.17 *Let* $p = (x, y)$ *be a point in* $\mathbb{R}^2$ *and let*

$$\gamma_p(t) = \begin{pmatrix} \sin t & \cos t \\ \cos t & \sin t \end{pmatrix} \begin{pmatrix} x \\ y \end{pmatrix}, \quad t \in \mathbb{R}$$

be a curve with initial point p *in* $\mathbb{R}^2$. *Find the velocity vector* v *in* $\mathbb{R}^2$. *Hence compute* $D_v f$, *where*

 (i) $f : \mathbb{R}^2 \to \mathbb{R}$ *is defined by* $f = x^2 y$.
 (ii) $f : \mathbb{R}^2 \to \mathbb{R}$ *is defined by* $f = e^x \cos y$.
 (iii) $f : \mathbb{R}^2 \to \mathbb{R}$ *is defined by* $f = x \cos y$.

Exercise 1.18 *Let* $p = (1, 0, 0)$ *be a point in* $\mathbb{R}^3$ *and let*

$$\gamma_p(t) = (e^t, \sin 2t, t), \quad t \in \mathbb{R}$$

be a curve with initial point p *in* $\mathbb{R}^3$. *Find the velocity vector* v *in* $\mathbb{R}^2$. *Hence compute* $D_v f$, *where*

 (i) $f : \mathbb{R}^2 \to \mathbb{R}$ *is defined by* $f = x^3 y$.
 (ii) $f : \mathbb{R}^2 \to \mathbb{R}$ *is defined by* $f = x e^z$.
 (iii) $f : \mathbb{R}^2 \to \mathbb{R}$ *is defined by* $f = y e^x \cos z$.

Answers

1.16 (i) 96   (ii) $4 \cdot 7^7$ (iii) $2e^2 (\cos 7 - 2 \sin 7)$
1.17 (i) $x^2 y$ (ii) $e^x (x \cos y + y \sin y)$ (iii) $x \cos y + xy \sin y$
1.18 (i) 2   (ii) 2   (iii) $2e$

1.3 Germ of a Function

Let us consider the set of all pairs (f, U) where $f : U \subset \mathbb{R}^n \to \mathbb{R}$ is a C^∞ function and U is the neighbourhood of a point p of $\mathbb{R}^n$. If (g, V) is another such pair, then we define an equivalence relation in this way:

(f, U) is equivalent to (g, U), symbolically $(f, U) \sim (g, V)$ if there exists an open set W such that $f = g|_W$, where $p \in W \subset U \cap V$. It can be shown that '$\sim$' is an equivalence relation. The equivalence class (f, U) is called the germ of f at $p \in U \subset \mathbb{R}^n$. We write it as $F(p)$. It can be shown that such $F(p)$ of $\mathbb{R}^n$ is

(i) an algebra over $\mathbb{R}$.
(ii) a module over $\mathbb{R}$,
where the defining relations are

$$\begin{cases} (f + g)(p) = f(p) + g(p), \\ (fg)(p) = f(p)g(p), \\ (\lambda f)(p) = \lambda f(p), \qquad \forall\, f, g \in F(p), \lambda \in \mathbb{R}. \end{cases} \tag{1.8}$$

The proof, of (i) and (ii) stated above, is beyond the scope of this book and hence it is left to the reader.

From (1.5), we now observe that

$$D_v(fg)(p) = (D_v f)g(p) + f(p)(D_v g). \tag{1.9}$$

Such an equation is also known as Leibnitz Product Rule.

Problem 1.12 Let $v = (x^3, -yz, 0)$ denote a vector in $\mathbb{R}^3$. For a fixed point $p = (x, y, z)$, compute $D_v(fg)$ where $f : \mathbb{R}^3 \to \mathbb{R}$ and $g : \mathbb{R}^3 \to \mathbb{R}$ are defined respectively by $f = xz$ and $g = y^2$.

Solution: We know from (1.9) that

$$D_v(fg)(p) = (D_v f)g(p) + f(p)D_v g. \tag{1.10}$$

Now

$$p + tv = (x, y, z) + t(x^3, -yz, 0) = (x + tx^3, y - tyz, z).$$

Thus

$$f(p + tv) = (x + tx^3)z = xz + tx^3 z.$$

Therefore

$$D_v f = \frac{d}{dt} f(p + tv)\big|_{t=0} = x^3 z,$$

$$g(p) = y^2, \quad f(p) = xz,$$

$$g(p + tv) = (y - tyz)^2.$$

Therefore

$$D_v g = \frac{d}{dt} g(p + tv)\big|_{t=0} = -2y^2 z.$$

Thus, from (1.10), we have

$$D_v (fg)(p) = x^3 y^2 z - 2xy^2 z^2.$$

Alternative

Here

$$D_v f = \sum v^i \frac{\partial f}{\partial x^i}(p) = x^3 z.$$

Also

$$D_v g = \sum v^i \frac{\partial g}{\partial x^i}(p) = -2y^2 z.$$

Using (1.10), we get the desired result.

Problem 1.13 *Compute $D_v(fg)$ where $f : \mathbb{R}^3 \to \mathbb{R}$ and $g : \mathbb{R}^3 \to \mathbb{R}$ are defined respectively as $f = xy^2 - yz^2$, $g = xe^y$ and $v = (-1, 2, 1)$ denotes a vector in $\mathbb{R}^3$, for a fixed $p = (2, -2, 1)$.*

Solution: Here

$$p + tv = (2 - t, -2 + 2t, 1 + t)$$

$$f(p) = 10$$

$$f(p + tv) = (2 - t)(2t - 2)^2 - 2(t - 1)(t + 1)^2.$$

Now

$$\frac{d}{dt} f(p + tv)\big|_{t=0} = D_v f = -18.$$

Again

$$g(p) = 2e^{-2}, \quad g(p + tv) = (2 - t)e^{2t-2}.$$

So

$$\frac{d}{dt} g(p + tv)\big|_{t=0} = D_v g = 3e^{-2}.$$

Thus

$$D_v (fg)(p) = -6e^{-2}.$$

Exercises

Exercise 1.19 *For any constant function C, prove that $D_v C = 0$.*

Exercise 1.20 *Let $v = (x, -y)$ denote a vector in $\mathbb{R}^2$. For a fixed point $p = (x, y)$, compute $D_v(fg)$ where $f : \mathbb{R}^2 \to \mathbb{R}$ and $g : \mathbb{R}^2 \to \mathbb{R}$ are defined respectively by:*

(i) $f = x^2 y$ and $g = e^x \cos y$
(ii) $f = e^x \sin y$ and $g = x \cos y$
(iii) $f = xe^y$ and $g = ye^x$.

Exercise 1.21 *Let $v = (1, -2, 1)$ denote a vector in $\mathbb{R}^3$. For a fixed point $p = (2, -2, 1)$, compute $D_v(fg)$, where $f : \mathbb{R}^3 \to \mathbb{R}$ and $g : \mathbb{R}^3 \to \mathbb{R}$ are defined respectively by:*

(i) $f = x^2 yz$ and $g = e^x \cos y$
(ii) $f = e^x \sin y$ and $g = xz \cos y$

Answers

(1.19) Use (1.9) and $\mathbb{R}$-linearity property.
(1.20) (i) $(x^2 y \cos y + x^3 y \cos y + x^2 y^2 \sin y)e^x$. **(ii)** $e^x(x^2 + x)\sin y \cos y - e^x xy \cos 2y$.
  (iii) $(x^2 y - xy^2)e^{x+y}$.
(1.21) (i) $16e^2(\sin 2 - 2\cos 2)$. **(ii)** $-e^2(5\cos 2 \sin 2 + 4 \cos 4)$.

1.4 Inverse Function Theorem

Suppose U be some open subset of the Euclidean space $\mathbb{R}^n$ and the non-linear mapping $F : U \to \mathbb{R}^n$ is continuously differentiable. Let $\tilde{x} \in U$. Suppose that, at the point the differential $F'(\tilde{x}) : \mathbb{R}^n \to \mathbb{R}^n$ is one-to-one and onto. This implies that the non-linear map F inherits local invertibility in the vicinity of the point $\tilde{x}$. Precisely, we can say that $\exists$ an open subset V of $\mathbb{R}^n$ such that $\tilde{x} \in V$ and an open subset W of $\mathbb{R}^n$ satisfying $F : V \to W$ is one-to-one and onto. Also, the inverse F^{-1} is continuously differentiable. This originates the notion of Inverse Function theorem.

Theorem 1.1 (Inverse Function Theorem of a single variable) *Let $U \subseteq \mathbb{R}$ be open and suppose that the function $F : U \to \mathbb{R}$ is a continuously differentiable function. Let $a \in U$ such that $f'(a) \neq 0$. Then there exists an open interval I containing the point a and an open interval J containing its image $f(a)$ such that the function $f : I \to J$ is one-to-one and onto. Moreover, the inverse function theorem $f^{-1} : J \to I$ is also continuously differentiable, and for a point y in J, if x is a point in I at which $f(x) = y$, then (Fig. 1.2)*

$$\left(f^{-1}\right)'(y) = \frac{1}{f'(x)}.$$

Fig. 1.2 The inverse
function theorem of a single
variable

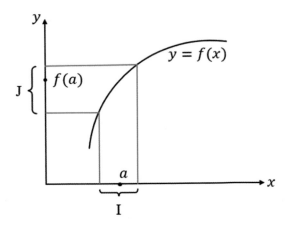

Proof Suppose $f'(a) > 0$. Since a is an interior point of U and the function f' : $U \to \mathbb{R}$ is continuous, therefore $\exists$ a real quantity $s > 0$ such that the closed interval $[a - s, a + s] \subset U$ and $f'(x) > 0$ for all points $x \in [a - s, a + s]$. By virtue of the Mean value theorem, we can say that the function $f : [a - s, a + s] \to \mathbb{R}$ is strictly increasing. In particular, $f : [a - s, a + s] \to \mathbb{R}$ is one-to-one. Furthermore, taking into consideration the Intermediate Value Theorem, if the point y lies between $f(a - s)$ and $f(a) + s$, $\exists x \in (a - s, a + s)$ with $f(x) = y$. Let us define $I = (a - s, a + s)$ and $J = (f(a) - s, f(a) + s) = (b - s, b + s)$ where $b = f(a)$. Then the function $f : I \to J$ is one-to-one and onto.

For the concluding part of the theorem, it follows from the Intermediate Value theorem that J is a neighbourhood of b. For $y \in J$, with $y \neq b$ define $x \equiv f^{-1}$ so that

$$\frac{f^{-1}(y) - f^{-1}(b)}{y - b} = \frac{1}{\frac{f(x) - f(a)}{x - a}}.$$

Since the inverse function $f^{-1} : J \to \mathbb{R}$ is continuous, therefore

$$\lim_{y \to b} x \equiv \lim_{y \to b} f^{-1}(y) = f^{-1}(b) \equiv a.$$

By the composition property for limits, the quotient property of limits, and the definition of the differentiability of $f : I \to J$ at a, it follows that

$$\lim_{y \to b} \frac{f^{-1}(y) - f^{-1}(b)}{y - b} = \lim_{y \to b} \frac{1}{\frac{f(x) - f(a)}{x - a}} = \frac{1}{f'(a)}.$$

Thus f^{-1} is differentiable at b, and its derivative at b is given by $\left(f^{-1}\right)'(y) = \frac{1}{f'(x)}$.

Theorem 1.2 (Inverse Function Theorem in the plane) *Let $U(\subseteq \mathbb{R}^2)$ open and suppose that the mapping $F : U \to \mathbb{R}^2$ is continuously differentiable. Let $(a, b) \in U$*

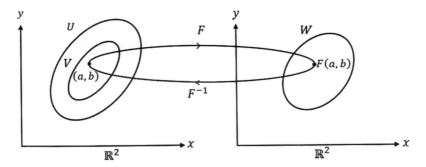

Fig. 1.3 The inverse function theorem of double variables

such that the derivative matrix $F'(a, b)$ be invertible. Then $\exists$ a neighbourhood V of (a, b) and a neighbourhood W of its image $F(a, b)$ such that $F : V \to W$ is one-to-one and onto. Moreover, the inverse mapping $F^{-1} : W \to V$ is also continuously differentiable, and for a point $(u, v) \in W$, if $(x, y) \in V$ such that $F(x, y) = (u, v)$, then the derivative matrix of the inverse mapping at the point (u, v) is given by the formula (Fig. 1.3)

$$(F^{-1})'(u, v) = F'(x, y)^{-1}.$$

Observe that in the proof of the last theorem, we used the Intermediate Value Theorem, a result that does not easily generalize to mappings whose image lies in the plane $\mathbb{R}^2$. An $n \times n$ matrix is invertible if and only if its determinant is non-zero, and when the matrix is invertible, there is a formula called Cramer's Rule for the inverse matrix. For 2×2 matrices, Cramer's Rule is clear by inspection. Indeed, for a 2×2 matrix $A = \begin{pmatrix} a_{11} & a_{12} \\ a_{21} & a_{22} \end{pmatrix}$, if $\det A \neq 0$ then $A^{-1} = \frac{1}{\det A} \begin{pmatrix} a_{22} & -a_{12} \\ -a_{21} & a_{11} \end{pmatrix}$. In particular, for the mapping $F : U \to \mathbb{R}^2$ in the statement of the Inverse Function Theorem in the Plane $F'(a, b)$ holds if and only if $\det F'(a, b) \neq 0$. If the mapping $F : U \to \mathbb{R}^2$ is represented in terms of component function as

$$F(x, y) = (F_1(x, y), F_2(x, y)), \quad (x, y) \in U,$$

then

$$F'(x, y) = \begin{pmatrix} \frac{\partial F_1}{\partial x}(x, y) & \frac{\partial F_1}{\partial y}(x, y) \\ \frac{\partial F_2}{\partial x}(x, y) & \frac{\partial F_2}{\partial y}(x, y) \end{pmatrix}.$$

So the assumption $\det F'(a, b) \neq 0$ is equivalent to

$$\frac{\partial F_1}{\partial x}(a, b)\frac{\partial F_2}{\partial y}(a, b) - \frac{\partial F_1}{\partial y}(a, b)\frac{\partial F_2}{\partial x}(a, b) \neq 0.$$

The above explicit formula for the inverse of a 2×2 matrix permits us to use formula $(F^{-1})'(u, v) = F'(x, y)^{-1}$ to compute the partial derivatives of the component functions of the inverse mapping $F^{-1} : W \to V$. Indeed, write the inverse mapping in component functions as

$$F^{-1}(u, v) = (g(u, v), h(u, v)), \quad (u, v) \in W,$$

such that

$$(F^{-1})'(u, v) = \begin{pmatrix} \frac{\partial g}{\partial u}(u, v) & \frac{\partial g}{\partial v}(u, v) \\ \frac{\partial h}{\partial u}(u, v) & \frac{\partial h}{\partial v}(u, v) \end{pmatrix}.$$

For a point $(u, v) \in W$, let $(x, y) \in V$ at which $u = F_1(x, y)$, $v = F_2(x, y)$. For notation, set $J(x, y) \equiv F'(x, y)$. Then, using the above computation of the inverse of a 2×2 matrix, it follows that formula $F^{-1}(u, v) = [F'(x, y)]^{-1}$ is equivalent to

$$\frac{\partial g}{\partial u}(u, v) = \frac{1}{J(x, y)} \cdot \frac{\partial F_2}{\partial y}(x, y)$$

$$\frac{\partial g}{\partial v}(u, v) = -\frac{1}{J(x, y)} \cdot \frac{\partial F_1}{\partial y}(x, y)$$

$$\frac{\partial h}{\partial u}(u, v) = -\frac{1}{J(x, y)} \cdot \frac{\partial F_2}{\partial x}(x, y)$$

$$\frac{\partial h}{\partial v}(u, v) = \frac{1}{J(x, y)} \cdot \frac{\partial F_1}{\partial x}(x, y).$$

Example 1.4 For a point $(x, y) \in \mathbb{R}^2$, let us define

$$F(x, y) = (\exp(x - y) + x^2 y + x(y - 1)^5, \ 1 + x^2 + x^4 + (xy)^5).$$

Since each of its component functions is continuously differentiable, therefore, the mapping $F : \mathbb{R}^2 \to \mathbb{R}^2$ is continuously differentiable. At the point $(a, b) = (1, 1)$, we have

$$F'(1, 1) = \begin{pmatrix} 3 & 0 \\ 11 & 5 \end{pmatrix}.$$

The determinant of $F'(1, 1)$ is non-zero. In view of the Inverse Function Theorem, $\exists$ neighbourhoods V of the $(1, 1)$ and W of its image $(2, 4)$ such that the mapping $F : U \to V$ is one-to-one and onto and that the inverse mapping $F^{-1} : V \to U$ is also continuously differentiable. Moreover, if the inverse is represented in components as $F^{-1}(u, v) = (g(u, v), h(u, v))$, then it follows that

$$\frac{\partial g}{\partial u}(2, 4) = \frac{1}{3}, \ \frac{\partial g}{\partial v}(2, 4) = 0, \ \frac{\partial h}{\partial u}(2, 4) = -\frac{11}{15}, \ \frac{\partial h}{\partial v}(2, 4) = \frac{1}{5}.$$

Example 1.5 Let us define the function $f : \mathbb{R}^2 \to \mathbb{R}^2$ by

$$f(x, y) = (\cos(x^2 + y), \sin(x^2 + y)), \ (x, y) \in \mathbb{R}^2.$$

Then f is continuously differentiable as each component function is also so. But $\nexists$ is any point at which the conclusion of the Inverse Function Theorem holds true. To see this, observe that if $(u, v) \in f(\mathbb{R}^2)$, then $u^2 + v^2 = 1$, a circle of radius 1 with centre at the origin. The image does not contain an open subset of the plane, so $\nexists$ any open sets U and V in $\mathbb{R}^2$ such that $f : U \to V$ is one-to-one and onto.

Example 1.6 Define the function $f : \mathbb{R}^2 \to \mathbb{R}^2$ by

$$f(x, y) = (x^2 - y^2, 2xy), \ (x, y) \in \mathbb{R}^2.$$

Since each of its component function is obviously continuously differentiable, therefore, f is also so. Consider $(a, b)(\neq (0, 0)) \in \mathbb{R}^2$. We have

$$f'(a, b) = \begin{pmatrix} 2a & -2b \\ 2b & 2a \end{pmatrix},$$

so $|f'(a, b)| = 4(a^2 + b^2) \neq (0, 0)$. Applying Inverse Function Theorem, it follows that there exist neighbourhoods U of (a, b) and V of $f(a, b)$ such that the mapping $f : U \to V$ is one-to-one and onto and has an inverse $f^{-1} : V \to U$ that also is continuously differentiable. Suppose $f^{-1}(\tilde{a}, \tilde{b}) = (g(\tilde{a}, \tilde{b}), h(\tilde{a}, \tilde{b}))$. Then, if we set $(\tilde{a}_\circ, \tilde{b}_\circ) = f(a, b)$, it follows that

$$\frac{\partial g}{\partial \tilde{a}}(\tilde{a}_\circ, \tilde{b}_\circ) = \frac{a}{2(a^2 + b^2)}, \quad \frac{\partial g}{\partial \tilde{b}}(\tilde{a}_\circ, \tilde{b}_\circ) = \frac{b}{2(a^2 + b^2)}$$
$$\frac{\partial h}{\partial \tilde{a}}(\tilde{a}_\circ, \tilde{b}_\circ) = \frac{-b}{2(a^2 + b^2)}, \quad \frac{\partial h}{\partial \tilde{b}}(\tilde{a}_\circ, \tilde{b}_\circ) = \frac{a}{2(a^2 + b^2)}.$$

But the assumptions of the Inverse Function Theorem fails to be true at the point $(0, 0)$, since

$$f'(0, 0) = \begin{pmatrix} 0 & 0 \\ 0 & 0 \end{pmatrix}.$$

Moreover, Inverse Function Theorem also fails at this point because if $f(x, y) = f(-x, -y)$ holds for every (x, y) in the plane, $\nexists$ neighbourhood of $(0, 0)$ on which the mapping f is one-to-one.

Problem 1.14 *The point $(1, e)$ lies on the graph of $y = xe^x$. Find an open set containing $y = e$, such that $\exists$ is a continuous function $x = g(y)$ defined on it, for which $x = g(y) \Rightarrow y = xe^x$ and $g(e) = 1$.*

Solution: Since $\dfrac{dy}{dx} = (1+x)e^x > 0$ on $(-1, \infty)$, the given function is injective when restricted to this interval and has range $(-e^{-1}, \infty)$, which is an open subset U of $\mathbb{R}$ containing e. Therefore, there is a continuous inverse g with domain U.

An Interpretation of the Inverse Function Theorem

Given two functions $F_1 : \mathbb{R}^2 \to \mathbb{R}$ and $F_2 : \mathbb{R}^2 \to \mathbb{R}$ and two numbers a_1 and a_2, consider the system of equations

$$F_1(x, y) = a_1; \quad F_2(x, y) = a_2.$$

A natural question arises whether there exist any solutions to this system of equations and, if there be any, then the solution is unique. If we define the mapping $F : \mathbb{R}^2 \to \mathbb{R}^2$ by $F(x, y) = \big(F_1(x, y), F_2(x, y)\big)$ for $(x, y) \in \mathbb{R}^2$, these two questions about the existence and uniqueness of the solutions of the system can be rephrased as questions about the image of the mapping $F : \mathbb{R}^2 \to \mathbb{R}^2$ and whether it has the property of being one-to-one. The following example shows how the Inverse Function Theorem provides information about systems of equations. Consider the system of equations

$$e^{x-y} + x^2 y + x(y-1)^5 = 2; \, 1 + x^2 + x^4 + (xy)^5 = 4.$$

Observe that the point $(x, y) = (1, 1)$ is a solution of this system. The mapping $F : \mathbb{R}^2 \to \mathbb{R}^2$ defined by

$$F(x, y) = \big(e^{x-y} + x^2 y + x(y-1)^5, 1 + x^2 + x^4 + (xy)^5\big).$$

for (x, y) in $\mathbb{R}^2$ is precisely the mapping considered in the Example 1.4. Referring to Example 1.4, we can say that $\exists$ a $\delta > 0$ and a neighbourhood U of the point $(1, 1)$ such that for any numbers a_1 and a_2 with $(a_1 - 2)^2 + (a_2 - 4)^2 < \delta^2$, the system of equations

$$e^{x-y} + x^2 y + x(y-1)^5 = a_1; \quad 1 + x^2 + x^4 + (xy)^5 = a_2$$

has exactly one solution.

Now we are going to state General Inverse Function Theorem.

Theorem 1.3 (Inverse Function Theorem on $\mathbb{R}^n$) *Let $U (\subseteq \mathbb{R}^n)$ be open and suppose that the mapping $F : U \to \mathbb{R}^n$ is continuously differentiable. Let $x_* \in U$ at which the derivative matrix $F'(x_*)$ is invertible. Then $\exists$ an open neighbourhood V of the point x_* and an open neighbourhood W of its image $F(x_*)$ such that the mapping $F : V \to W$ is one-to-one and onto. Moreover, the inverse mapping $F^{-1} : W \to V$*

is also continuously differentiable, and for a point $y \in W$, if x is the point in V such that $F(x) = y$, then

$$(F^{-1})'(y) = F'(F^{-1}(y))^{-1}.$$

Before we proceed with the proof of the Inverse Function Theorem for n-variables, let us prove the following lemma:

Lemma 1.1 *Suppose $f : O(\subset \mathbb{R}^n) \to \mathbb{R}^m$ is differentiable in a convex open set O and there exists $M \in \mathbb{R}$ such that $|f'(x)| \le M, \ \forall \ x \in O$. Then*

$$||f(b) - f(a)|| \le M||b - a||, \ \forall \ a, b \in O.$$

Proof Let us fix a, b in O and define $\gamma : \mathbb{R} \to \mathbb{R}^n, \ \forall \ t \in \mathbb{R}$. Since O is convex, $\gamma(t) \in E$ provided $t \in [0, 1]$. Let us set $g(t) = (f \circ \gamma)(t)$. Then

$$g'(t) = f'(\gamma(t))\gamma'(t) = f'(\gamma(t))(b - a),$$

which implies

$$|g'(t)| \le |f'(\gamma(t))| ||b - a|| \le M||b - a||, \ \forall \ t \in [0, 1].$$

This completes the proof of the lemma.

Proof of the Main Theorem:
Since F' is continuous at x_*, therefore, for a preassigned $\epsilon > 0$ there exists an open neighbourhood $V \subset U$ of x_* such that

$$x \in V \Rightarrow |F'(x) - F'(x_*)| < \epsilon. \tag{1.11}$$

Let us choose $\epsilon = \dfrac{1}{2} \dfrac{1}{|F'(x_*)^{-1}|}$. Then the preceding equation becomes

$$|F'(x) - F'(x_*)| < \frac{1}{|F'(x_*)^{-1}|} \Rightarrow |F'(x_*)^{-1}| \cdot |F'(x) - F'(x_*)| < \frac{1}{2}. \tag{1.12}$$

Then we see that for $x \in V$, $F'(x)$ is invertible. Now, for any $y \in \mathbb{R}^n$, let us define a function $\psi_y : V \to \mathbb{R}^n$ by

$$\psi_y(x) = x + F'(x_*)^{-1}(y - F(x)) = F'(x_*)^{-1}(F'(x_*) + y - F(x)). \tag{1.13}$$

Then

$$x \text{ is a fixed point of } \psi_y \Leftrightarrow y = F(x). \tag{1.14}$$

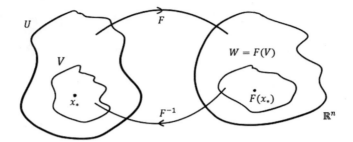

Fig. 1.4 Inverse function theorem on $\mathbb{R}^n$

Now as a consequence of the chain rule the composition function ψ_y has

$$\psi_y'(x) = F'(x_*)^{-1}\big(F'(x_*) - F'(x)\big).$$

Combining (1.12) with the last equation, we obtain

$$|\psi_y'(x)| < \frac{1}{2}, \quad x \in V. \tag{1.15}$$

Hence, by virtue of Lemma 1.1, it follows that

$$\|\psi_y(x_1) - \psi_y(x_2))\| \le \|x_1 - x_2\|, \ \forall\, x_1, x_2 \in V. \tag{1.16}$$

Thus, ψ_y has at most one fixed point in V so that by (1.15), $F(x) = y$ holds for at most one $x \in V$. This proves F is injective on V.

Next, consider $F(V) = W$. Then $F : V \to W$ is an injective map. Therefore, there exists an inverse map $F^{-1} : W \to V$ as illustrated in the figure below (Fig. 1.4):

In order to complete the proof of the first part of the theorem, it only remains to show W is open. Choose $w_o \in W$ be arbitrary. Then for some $x_o \in V$, $F(x_o) = w_o$. Let $\epsilon > 0$ be sufficiently small enough such that $\|x - x_o\| \le \epsilon \Rightarrow x_o \in V$, so that

$$B_\epsilon(x_o) = \{x \in \mathbb{R}^n : \|x - x_o\| \le \epsilon\} \subset V.$$

To show W is open, it suffices to show that

$$y \in \mathbb{R}^n, \ \|y - w_o\| < \frac{\epsilon}{2|F'(x_*)^{-1}|} \Rightarrow y \in W.$$

In order to prove this, suppose $\|y - w_o\| < \frac{\epsilon}{2|F'(x_*)^{-1}|}$. Then for any $x \in B_\epsilon(x_o)$, we find

$$\|\psi_y(x) - x_o\| \le \|\psi_y(x) - \psi_y(x_o)\| + \|\psi_y(x_o) - x_o\|$$

$$\le \frac{1}{2}\|x - x_o\| + \|F'(x_*)^{-1}(y - F(x_o))\|, \text{ by } (1.13), (1.16)$$

$$\le \frac{\epsilon}{2} + |F'(x_*)^{-1}| \|y - w_o\|$$

$$\le \frac{\epsilon}{2} + |F'(x_*)^{-1}| \frac{\epsilon}{2|F'(x_*)^{-1}|}$$

$$\le \epsilon,$$

which proves $\psi_y(x) \in B_\epsilon(x_o)$. Consequently, by virtue of (1.16), $\psi_y : B_\epsilon(x_o) \to B_\epsilon(x_o)$ is a contraction map, where $B_\epsilon(x_o)$ is closed in $\mathbb{R}^n$. We can invoke contraction principle in $\mathbb{R}^n$, to conclude that ψ_y has a unique fixed point in $B_\epsilon(x_o) \subset V$. Thus, by (1.14), we have $y \in F(B_\epsilon(x_o)) \subset F(V) = W$. This completes the argument that W is open.

Furthermore, for any $\epsilon > 0$

$$\|y - w_o\| < \frac{\epsilon}{2|F'(x_*)^{-1}|}, \quad y \in W \Rightarrow y \in F(B_\epsilon(x_o))$$

$$\Rightarrow y = F(x) \text{ for some } x \in B_\epsilon(x_o)$$

$$\Rightarrow F^{-1}(y) = B_\epsilon(x_o)$$

$$\Rightarrow \|F^{-1}(y) - x_o\| \le \epsilon,$$

which proves the continuity of the inverse map F^{-1}.

To prove the second part of the theorem, let us proceed as follows:

Here for sufficiently small h, we have

$$F\big(F^{-1}(y) + h\big) - F(F^{-1}(y)) = F'(F^{-1}(y))(h) + \|h\|R(h), \qquad (1.17)$$

where $R(h)$ being the remainder term and $R(h) \to 0$ as $\|h\| \to 0$. Note that, $F^{-1}(y) \in V \Rightarrow F'(F^{-1}(y))$ is invertible. Since F^{-1} is continuous, therefore, for sufficiently small k, suppose $h = F^{-1}(y + k) - F^{-1}(y)$. Then for every such k, we find

$$F\big(F^{-1}(y + k)\big) - F(F^{-1}(y)) = F'(F^{-1}(y))(h) + \|h\|R(h).$$

Applying $F'(F^{-1}(y))^{-1}$ to both sides of the last equality, we get

$$h = F'(F^{-1}(y))^{-1}(k) - \|h\|F'(F^{-1}(y))^{-1}(R(h)), \qquad (1.18)$$

which is the same as

$$F^{-1}(y + k) - F^{-1}(y) = F'(F^{-1}(y))^{-1}(k) - \|h\|F'(F^{-1}(y))^{-1}(R(h)).$$

We claim the existence of $(F^{-1})'(y)$ and $(F^{-1})'(y) = F'(F^{-1}(y))$. For this, it suffices to show that

$$\frac{||h||}{||k||} F'(F^{-1}(y))(R(h)) \to 0 \text{ as } ||k|| \to 0.$$

By continuity of F^{-1}, we have $||k|| \to 0 \Rightarrow ||h|| \to 0$. So it is sufficient to prove $\frac{||h||}{||k||}$ remains bounded. Now (1.18) yields

$$||h|| \leq |F'(F^{-1}(y))|\, ||k|| + ||h||\, |F'(F^{-1}(y))|\, ||R(h)||,$$

which further implies

$$\frac{||h||}{||k||} \left(1 - |F'(F^{-1}(y))|\, ||R(h)||\right) \leq |F'(F^{-1}(y))|.$$

Since $||h|| \to 0 \Rightarrow R(h) \to 0$, it follows that $\dfrac{||h||}{||k||}$ remains bounded. Thus, for every $y \in W$, we prove the existence of $(F^{-1})'(y)$ and $(F^{-1})'(y) = F'(F^{-1}(y))$. Furthermore, the equality shows that $(F^{-1})'$ is the composition of F^{-1}, F' and the inversion maps. All these are continuous, and therefore, $(F^{-1})'$ is continuous. This establishes the last part of the theorem.

Remark 1.3 1. In summary, **F is locally invertible at x_* with a continuously differentiable local inverse** or **F has a continuously differentiable local inverse at x_***. The term **local inverse** here refers to the function F^{-1}.

2. In the proof of the above theorem, we have used the notion **norm of a linear mapping** and the **Contraction Principle**. For details, the reader may refer to any standard book on Linear Algebra and Multivariable Analysis.

3. The benefit behind using the contraction principle in the proof of Inverse Function Theorem is that, it can be extended to the condition when $\mathbb{R}^n$ is replaced by an infinite-dimensional space. Alternatively, it is possible to prove the theorem using the compactness of a closed ball in $\mathbb{R}^n$, where the above benefit fails to be true.

Problem 1.15 *Let U and V be open subsets of $\mathbb{R}^n$ and let $F : U \to V$ be continuously differentiable and bijective, so that the inverse map $F^{-1} : V \to U$ exists. Suppose $F'(x)$ is invertible for every $x \in U$. Show that $(F^{-1})'$ exists on the entire given set V.*

Solution: Let $q \in V$. Then $q = F(p)$ for some $p \in U$. Since $F'(p)$ is invertible, in view of the inverse function theorem $\exists$ open sets $U_1 \subset U$ and $V_1 \subset \mathbb{R}^n$ such that $p \in U_1$ and $F(U_1) = V_1$ and F has a differentiable local inverse on V_1. But then $q = F(p) \in V_1$ and the local inverse is, therefore, differentiable at q. However, F^{-1} has to agree with the local inverse, and therefore, differentiable at q.

Problem 1.16 *If $F : E \to \mathbb{R}^n$ is a continuously differentiable mapping of an open set $E \subset \mathbb{R}^n$ and if $f'(x)$ is invertible for every $x \in E$, then prove that f is an open mapping of E into R^n.*

Solution: Let $U \subset E$ be open and $b \in F(U)$. Then $q = F(p)$ for some $p \in U$. Since $F'(p)$ is invertible, the inverse function theorem yields open sets $U_1 \subset U$ and $V \subset \mathbb{R}^n$ such that $p \in U_1$ and $f(U_1) = V$. But then $F(p) \in V \subset f(U)$. Since $q = F(p)$ and V is open, implies $f(U)$ is open.

Remark 1.4 Here $f : E \to \mathbb{R}^n$ is an open mapping means that f maps every open subset of E into an open subset of R^n.

Problem 1.17 *Let U be an open subset of $\mathbb{R}^n$ and let $f : U \to \mathbb{R}^n$ be a continuously differentiable map such that $f'(x)$ is invertible for every $x \in U$. Suppose V is an open subset of U such that its closure $\bar{V}$ is bounded and contained in U, and f is injective on the closure. Show that the image $f(\bar{V})$ is the closure of an open set.*

Solution: It is trivial to show that $f(\bar{V}) \subset \overline{f(V)}$. To prove the reverse inclusion, consider any $y \in \overline{f(V)}$. Then $\exists$ a sequence x_n in V such that $f(x_n) \to y$. Since $\bar{V}$ is bounded (hence compact), $x_n \to x(\in \bar{V})$ when x_n is replaced by a suitable subsequence. Since f is continuous, it follows that $f(x_n) \to f(x)$, so that $y = f(x)$. Since $x \in \bar{V}$, we have $y \in f(\bar{V})$. So $\overline{f(V)} \subset f(\bar{V})$, and hence $\overline{f(V)} = f(\bar{V})$. By virtue of the last problem, note that the set $f(V)$, of which $\overline{f(V)}$ is the closure, is an open set.

Exercises

Exercise 1.22 *Define the function $F : \mathfrak{P} \to \mathbb{R}^2$, where $\mathfrak{P} = \{(x_1, x_2) \in \mathbb{R}^2 : x_1 \neq 0\}$, by*

$$y_1 = F_1(x_1, x_2) = x_1 \cos x_2, \ y_2 = F_2(x_1, x_2) = x_1 \sin x_2, \ x_1 \neq 0.$$

At what points $(x_1, x_2) \in \mathfrak{P}$ does the Inverse Function Theorem apply?

Exercise 1.23 *Consider the equation $\dfrac{x^2}{8} + \dfrac{y^2}{8} = 1$, $(x, y) \in \mathbb{R}^2$.*

(i) *Explicitly define the function $g : I \to \mathbb{R}$ that has the property that in the neighbourhood of the solution $(2, 3)$, all the solutions are of the form $(x, g(x))$ for $x \in I$ and check that*
(ii) *Explicitly define the function $h : J \to \mathbb{R}$ that has the property that in a neighbourhood of the solution $(2, 3)$, all the solutions are of the form $(h(y), y)$ for y in J.*

Exercise 1.24 *Let $f : \mathbb{R} \to \mathbb{R}$ such that $f(x, y) = (x^2 + y, 2x + y^2)$. Find f' and determine the values of (x, y) for which f is NOT invertible. Given that f is invertible at $(0, 0)$, let g be its inverse. Find $g'(0, 0)$.*

1.5   Implicit Function Theorem

Let U be an open subset of the plane $\mathbb{R}^2$ and $f : U \to \mathbb{R}$ is continuously differentiable function. In general, the solution set of the equation is not the graph of a function expressing y as a function of x. Hence, the solution set is a very complicated subset of the plane. However, if the point (a, b) is a solution of this equation and $\dfrac{\partial f}{\partial y}(a, b) \neq 0$, then $\exists$ a neighbourhood of the point (a, b) with the property that the solutions of the above equation that are in this neighbourhood make up the graph of a continuously differentiable function $g : I \to \mathbb{R}$, where I is an open interval about a. Moreover, the derivative of the implicitly defined function $g : I \to \mathbb{R}$ can be computed in terms of the partial derivatives of the function $f : U \to \mathbb{R}$. This concept is called Dini's Theorem. It has an extension, called the General Implicit Function Theorem, that provides a similar local description of the set of solutions of an equation of the form $F(u) = 0$, $u \in U$, where U is an open subset of Euclidean space $\mathbb{R}^{n+k}$ and the mapping $F : U \to \mathbb{R}^{n+k}$ is continuously differentiable.

Theorem 1.4 (Dini's Theorem) *Let $U \subseteq \mathbb{R}^2$ be open. Suppose that the function $f : U \to \mathbb{R}$ is continuously differentiable. Let $(a, b) \in U$ such that $f(a, b) = 0$ and $\dfrac{\partial f}{\partial y}(a, b) \neq 0$. Then $\exists$ a positive real quantity r and a continuously differentiable function $g : I \to \mathbb{R}$, where $I \equiv (a - r, a + r)$ such that*

$$f(x, g(x)) = 0, \ \forall \, x \in I \tag{1.19}$$

and

$$\text{whenever } |x - a| < r, \ |y - b| < r \text{ and } f(a, b) = 0, \ \text{then } y = g(x). \tag{1.20}$$

Moreover,

$$\frac{\partial f}{\partial x}(x, g(x)) + \frac{\partial f}{\partial y}(x, g(x))g'(x) = 0, \ \forall x \in I. \tag{1.21}$$

Proof Suppose $\dfrac{\partial f}{\partial y}(a, b) > 0$. Since U is open and the function $\dfrac{\partial f}{\partial y} : U \to \mathbb{R}$ is continuous and positive at the point (a, b), $\exists$ positive numbers m and n such that the rectangular region $R = [a - m, a + m] \times [b - m, b + m] \subset U$ and

$$\frac{\partial f}{\partial y}(x, y) \geq n \ \forall \, (x, y) \in R. \tag{1.22}$$

With the aid of the Mean Value Theorem for real-valued functions, if $|x - a| \leq m$ & $b - m \leq y_1 < y_2 \leq b + m$ holds, then

$$f(x, y_1) < f(x, y_2). \tag{1.23}$$

In particular, $f(a, b) = 0 \Rightarrow f(a, b - m) < 0 < f(a, b + m)$. Since $f : U \to \mathbb{R}$ is continuously differentiable, therefore, f is continuous. Then $\exists$ a positive real quantity $r(< m)$ such that

$$f(x, b - m) < 0 < f(x, b + m), \ \forall \, x \in I \equiv (a - r, a + r).$$

Let $x \in I$. Since $f(x, b - m) < 0$ and $f(x, b + m) > 0$, by virtue of the Intermediate Value Theorem, $\exists \, y \in (b - m, b + m)$ at which $f(x, y) = 0$, and (1.23) implies that $\exists$ only one such point, say $g(x)$. This defines a function $g : I \to \mathbb{R}$ having properties (1.19) and (1.20).

Our claim: $g : I \to \mathbb{R}$ is continuously differentiable and that the differentiation formula (1.21) holds at the point a. Indeed, let $a + h \in I$. Then by definition, $f(a + h, g(a + h)) = 0$ and $f(a, g(a)) = 0$. Hence, $f(a + h, g(a + h)) - f(a, g(a)) = 0$. Considering the Mean Value Theorem for scalar functions of two real variables, $\exists$ some points on the segment between the points $(a, g(a))$ and $(a + h, g(a + h))$, which we label $q(h)$, at which

$$f(a + h, g(a + h)) - f(a, g(a)) = \frac{\partial f}{\partial x}(q(h))h + \frac{\partial f}{\partial y}(q(h))[g(a + h) - g(a)].$$

This implies,

$$\frac{\partial f}{\partial x}(q(h))h + \frac{\partial f}{\partial y}(q(h))[g(a + h) - g(a)] = 0.$$

Thus

$$g(a + h) - g(a) = -\frac{\frac{\partial f}{\partial x}(q(h))}{\frac{\partial f}{\partial y}(q(h))} h. \qquad (1.24)$$

Since the function $\dfrac{\partial f}{\partial x} : I \to \mathbb{R}$ is continuous and the closed square R is a sequentially compact subset of the plane $\mathbb{R}^2$, the Extreme Value Theorem guarantees the existence of a positive number M, such that, for every $(x, y) \in R$

$$\frac{\partial f}{\partial x}(x, y) \geq M.$$

Combining the inequality (1.22) with the foregoing one, it follows from (1.24) that

$$|g(a + h) - g(a)| \leq \frac{M}{n}|h|, \ a + h \in I.$$

Hence, the function $g : I \to \mathbb{R}$ is continuous at the point a. Since the point $q(h)$ lies on the segment between the points $(a, g(a))$ and $(a + h, g(a + h))$, we conclude that

$$\lim_{h \to 0} q(h) = (a, b).$$

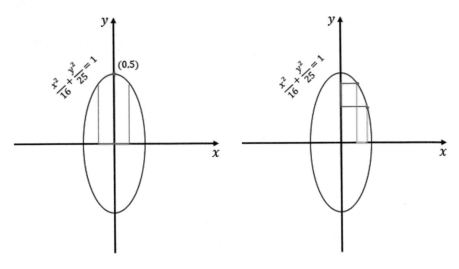

Fig. 1.5 Implicit Function Theorem

Dividing (1.24) by h and using the continuity of the first-order partial derivatives of $f : U \to \mathbb{R}$ at the point (a, b), it follows that

$$\lim_{h \to 0} \frac{g(a + h) - g(a)}{h} = -\frac{\frac{\partial f}{\partial x}(a, b)}{\frac{\partial f}{\partial y}(a, b)},$$

which means that g is differentiable at a and formula (1.21) holds at a. But any other point $x \in I$ satisfies the same assumptions as does the point a, and hence (1.21) holds at all points in I (Fig. 1.5).

Example 1.7 Let us consider the equation $\dfrac{x^2}{16} + \dfrac{y^2}{25} = 1$, $(x, y) \in \mathbb{R}^2$. The set of solutions of the given equation consists of points in $\mathbb{R}^2$ lying on an ellipse with $(0, 0)$ as its centre. Let us begin with the solution $(0, 5)$. Then for any r lying between 0 and 4, define an interval $I = (-r, r)$ and define the function $G : I \to \mathbb{R}$ by $G(x) = 5\sqrt{1 - \dfrac{x^2}{4^2}}$, $x \in I$. Then there exists a neighbourhood of $(0, 5)$ having the property that the set of solutions of the given equation in this neighbourhood consists of points of the form $(x, G(x))$, $\forall x \in I$.

Next, let us consider the second component of $(0, 5)$. Here, it is not possible to find a neighbourhood J of 5, a function $H : J \to \mathbb{R}$, and a neighbourhood of $(0, 5)$ in which the set of solutions of the given equation consists of the points of the form $(H(y), y)$, $\forall y \in J$. At every other vertices of the ellipse, it is possible to find a neighbourhood of the vertex in which the set of solutions of the given equation has a similar description. On the other hand, at (a, b) that is not a vertex of the ellipse, is a neighbourhood of (a, b) the set of solutions of the given equation determines both x as a function y and vice-versa.

Example 1.8 Consider the equation

$$\cos(x + y) + \exp(y + x^2) + 3x + (xy)^5 - 2 = 0, \quad (x, y) \in \mathbb{R}^2. \tag{1.25}$$

Let us define

$$f(x, y) = \cos(x + y) + \exp(y + x^2) + 3x + (xy)^5 - 2.$$

Then (x, y) is a solution of (1.25) if and only if $f(x, y) = 0$. Note that $(0, 0)$ is a solution of (1.25) and that

$$\frac{\partial f}{\partial x}(0, 0) = 3, \quad \frac{\partial f}{\partial y}(0, 0) = 1.$$

Here f is a continuously differentiable function. So taking advantage of Dini's Theorem, we find a positive number r and a continuously differentiable function $g : I \to \mathbb{R}$, where I is the open interval $(-r, r)$, such that

$$\cos(x + g(x)) + \exp^{g(x)+x^2} + 3x + x^5(g(x))^5 - 2 = 0, \quad \forall x \in I.$$

Moreover, if (x, y) is a solution of (1.25) with $|x| < r$ and $|y| < r$, then $y = g(x)$. Finally, $g'(0)$ is determined by the formula

$$\frac{\partial f}{\partial x}(0, 0) + \frac{\partial f}{\partial y}(0, 0)g'(0) = 0 \Rightarrow g'(0) = -3.$$

Example 1.9 Let us consider the equation $x^2 - y^2 = 0$, $(x, y) \in \mathbb{R}^2$. The set of solutions of the given equation consists of points in $\mathbb{R}^2$ lie on the line $x = y$ or $x = -y$. At each solution $(a, b) \neq (0, 0)$ of the given equation, there exists a neighbourhood of (a, b) in which the set of solutions of the given equation determines both x as a function of y and vice-versa. The origin $(0, 0)$ is a solution of the given equation, but $\nexists$ any neighbourhood of the origin in which the set of solutions coincides with the graph of a function expressing one of the components of (x, y) as a function of the other one.

Problem 1.18 *Show that the system of equations*

$$3x_1 + x_2 - x_3 - u^3 = 0$$
$$x_1 - x_2 + 2x_3 + u = 0$$
$$2x_1 + 2x_2 - 3x_3 + 2u = 0,$$

cannot be solved for x_1, x_2, x_3 in terms of u.

Solution: Let

$$f(x_1, x_2, x_3, u) = 3x_1 + x_2 - x_3 - u^3$$
$$g(x_1, x_2, x_3, u) = x_1 - x_2 + 2x_3 + u$$
$$h(x_1, x_2, x_3, u) = 2x_1 + 2x_2 - 3x_3 + 2u.$$

Then,

$$\Delta = \begin{vmatrix} \frac{\partial f}{\partial x_1} & \frac{\partial f}{\partial x_2} & \frac{\partial f}{\partial u} \\ \frac{\partial g}{\partial x_1} & \frac{\partial g}{\partial x_2} & \frac{\partial g}{\partial u} \\ \frac{\partial h}{\partial x_1} & \frac{\partial h}{\partial x_2} & \frac{\partial h}{\partial u} \end{vmatrix} = \begin{vmatrix} 3 & 1 & -3u^2 \\ 1 & -1 & 1 \\ 2 & 2 & 2 \end{vmatrix} = -12 - 12u^2,$$

which can never be 0. If there were to exist a solution for x_1, x_2, x_3 valid on some interval (in which u varies), then the fact that $(2x_1 + 2x_2 - 3x_3) = (3x_1 + x_2 - x_3) - (x_1 - x_2 + 2x_3)$ would imply that $-2u = u^3 + u$ on that interval, which is impossible.

Implicit Function Theorem on $\mathbb{R}^n$

Let m and n be positive integers. Suppose U be an open subset of $\mathbb{R}^{m+n}$, and and that the mapping $F : U \to \mathbb{R}^{m+n}$ is continuously differentiable. Consider the equation

$$F(u) = 0, \ u \in U.$$

In the case where $m = 1$ and $n = 1$, we already considered this equation in Dini's Theorem. The object of this theorem is to state the General Implicit Function Theorem, an extension of Dini's Theorem to more general equations of the form $F(u) = 0, \ u \in U$. In order to emphasize the analogy between the general case and the case where $m = 1$ and $n = 1$, it is useful to introduce the following notation: For a point $u \in \mathbb{R}^{m+n}$, we separate the first m components of u from the last n components and label them as follows:

$$u = (x, y) = (x_1, x_2, \ldots, x_m, y_1, y_2, \ldots, y_n), \ (x, y) \in U.$$

If the mapping $F : U \to \mathbb{R}^n$ is written in terms of its component functions, $F = (F_1, F_2, \ldots, F_n)$, this equation, in turn, can be expressed as the following system of n non-linear scalar equations in $m + n$ scalar unknowns:

$$F_1(x_1, x_2, \ldots, x_m, y_1, y_2, \ldots, y_n) = 0$$

$$\vdots$$

$$F_i(x_1, x_2, \ldots, x_m, y_1, y_2, \ldots, y_n) = 0$$

$$\vdots$$

$$F_n(x_1, x_2, \ldots, x_m, y_1, y_2, \ldots, y_n) = 0.$$

where $(x_1, x_2, \ldots, x_m, y_1, y_2, \ldots, y_n)$, $(x, y) \in U$.

Now we are in a position to state the General Implicit Function Theorem, a direct generalization of Dini's Theorem.

Statement 1.1 Implicit Function Theorem on $\mathbb{R}^n$: *Let m and n be positive integers. Suppose U be an open subset of $\mathbb{R}^{n+m} (\equiv \mathbb{R}^n \times \mathbb{R}^m)$ and that the mapping $F : U \to \mathbb{R}^n$ is continuously differentiable. At the point $(x_0, y_0) \in U$, suppose that $F(x_0, y_0) = 0$. Let $T_1 : \mathbb{R}^n \to \mathbb{R}^n$ and $T_2 : \mathbb{R}^m \to \mathbb{R}^n$ be two linear maps defined by*

$$T_1(h_1) = F'(x_0, y_0)(h_1, 0), \quad T_2(h_2) = F'(x_0, y_0)(0, h_2);$$

so that $F'(x_0, y_0)(h_1, h_2) = T_1(h_1) + T_2(h_2)$ for all $h_1 \in \mathbb{R}^n$, $h_2 \in \mathbb{R}^m$. Suppose T_1 is invertible. Then

(a) *there exists open sets $V_1 \subset U$, $V_2 \subset \mathbb{R}^m$ with $(x_0, y_0) \in V_1$, $y_0 \in V_2$ and a unique map $G : V_2 \to \mathbb{R}^n$ such that*

$$(G(y), y) \in V_1, \quad F(G(y), y) = 0 \,\forall\, y \in V_2;$$

(b) *for every $(x, y) \in V_1$ such that $F(x, y) = 0$, we have $y \in V_2$ and $x = G(y)$;*

(c) *furthermore, G is continuously differentiable and $G(y_0) = x_0$, $G'(x_0, y_0) = -T_1^{-1}T_2$.*

***Proof* (a)** Let us define a map $f : U \to \mathbb{R}^{n+m}$ by $f(x, y) = (F(x, y), y)$. Then f is differentiable on U and

$$f'(x, y)(h_1, h_2) = (F'(x, y)(h_1, h_2), h_2), \quad \forall\, (h_1, h_2) \in \mathbb{R}^{n+m}.$$

This implies f is continuously differentiable and it follows that

$$f'(x_0, y_0)(h_1, h_2) = (T_1(h_1) + T_2(h_2), h_2), \quad (h_1, h_2) \in \mathbb{R}^{n+m}.$$

Since T_1 is invertible, therefore

$$f'(x_0, y_0)(h_1, h_2) = 0 \Rightarrow (T_1(h_1) + T_2(h_2), h_2) = 0 \Rightarrow T_1(h_1) + T_2(h_2) = 0, \ h_2 = 0$$
$$\Rightarrow T_1(h_1) = 0, \ h_2 = 0$$
$$\Rightarrow h_1 = 0, \ h_2 = 0.$$

This shows $f'(x_0, y_0)$ is injective and so is also surjective, hence invertible. So by the virtue of Inverse Function Theorem, there exists open sets $V_1 \subset U$, $V_3 \subset \mathbb{R}^{n+m}$ such that $(x_0, y_0) \in V_1$, $f : V_1 \to V_3$ is injective and the inverse map $f^{-1} :$ $V_3 \to V_1$ is continuously differentiable. Moreover,

$$f(x_0, y_0) = (F(x_0, y_0), y_0) = (0, y_0).$$

So $(0, y_0) \in V_3$. Suppose $V_2 = \{y \in \mathbb{R}^m : (0, y) \in V_3\}$. Then $y_0 \in V_2$ and V_2 is open. For any $y \in V_2$, we have $(0, y) \in V_3$ and hence there exists $(x, z) \in V_1$ such that $f(x, z) = (0, y)$. But $f(x, z) = (F(x, z), z)$. Therefore, $(F(x, z), z) = (0, y)$, so that $y = z$ and $F(x, y) = 0$. If $F(\tilde{x}, y) = 0$ with $(\tilde{x}, y) \in V_1$, then $(F(\tilde{x}, y), y) = (0, y)$, i.e.,

$$F(\tilde{x}, y) = (0, y) = f(x, y).$$

But f is injective on V_1. So $\tilde{x} = x$. Hence, there exists a unique x for which $F(x, y) = 0$ and $(x, y) \in V_1$. Taking x as $G(y)$, (a) is established.

(b) Let $(x, y) \in V_1$ and $F(x, y) = 0$. Then $f(x, y) \in V_3$. But $f(x, y) = (F(x, y), y)$ and $F(x, y) = 0$. This means $(0, y) \in V_3$, so that $y \in V_2$. By definition of G above, $G(y)$ is the unique η such that $F(\eta, y) = 0$ and $(\eta, y) \in V_1$. This implies $x = \eta = G(y)$

(c) Since $F(x_0, y_0) = 0$, $(x_0, y_0) \in V_1$ and $y_0 \in V_2$, therefore, $G(y_0) = x_0$. Now, for any $y \in V_2$, we have $F(G(y), y) = 0$ by (a), so that $f(G(y), y) = (0, y)$, from which it follows that $(G(y), y) = f^{-1}(0, y)$. Thus, G is the composition of the maps

$$y \mapsto (0, y), \quad (x, y) \mapsto f^{-1}(x, y), \quad (x, y) \mapsto x,$$

where the first and third are linear maps, while the second is continuously differentiable. It follows that G is continuously differentiable. Since $F(G(y), y) = 0 \ \forall \ y \in V_2$, the mapping $y \mapsto F(G(y), y)$ must have derivative 0 everywhere. On the other hand, we find that the derivative of the mapping $y \mapsto F(G(y), y)$ at y_0 maps $h_2 \in \mathbb{R}^m$ into

$$F'(G(y_0), y_0)(G'(y_0), y_0) = F'(x_0, y_0)(G'(y_0)h_2, h_2), \quad \text{because } G(y_0) = x_0$$
$$= T_1 G'(y_0)h_2 + T_2 h_2, \quad \text{by hypothesis.}$$

Since $F'(G(y_0), y_0) = 0$, then $G'(y_0)h_2 = -T_1^{-1}T_2 h_2$ for all $h_2 \in \mathbb{R}^m$. This completes the proof of (c).

Remark 1.5 The implicit function theorem above provides a sufficient condition in order that a continuous solution G of $F(x, y) = 0$ for x in terms of y satisfying the requirement that $G(y_o) = x_o$ should exist and be unique. However, our theorem does not explicitly mention the word **solution**.

Example 1.10 Consider the system of equations

$$\ln(7 + x_2^2 + x_3^2) + x_1 x_3 + e^{x_1 + x_4} + 7 = 0,$$
$$x_1^3 \exp\{\cos(x_2^2 + x_4^2)\} + x_1 + 2x_4 + (x_2 + x_1 + x_4)^4 = 0,$$

where $(x_1, x_2, x_3, x_4) \in \mathbb{R}^4$. Note that the point $(0, 0, 0, 0)$ is a solution of this system of equations. For a point $(x_1, x_2, x_3, x_4) \in \mathbb{R}^4$, let us define

$$F(x_1, x_2, x_3, x_4) = \left(\ln(7 + x_2^2 + x_3^2) + x_1 x_3 + e^{x_1 + x_4} + 7, \; x_1^3 \exp\{\cos(x_2^2 + x_4^2)\} \right.$$
$$\left. + x_1 + 2x_4 + (x_2 + x_1 + x_4)^4 \right).$$

Here $F(x_1, x_2, x_3, x_4) = (F_1(x_1, x_2, x_3, x_4), \; F_2(x_1, x_2, x_3, x_4))$, where

$$F_1(x_1, x_2, x_3, x_4) \equiv \ln(7 + x_2^2 + x_3^2) + x_1 x_3 + e^{x_1 + x_4} + 7 = 0,$$
$$F_2(x_1, x_2, x_3, x_4) \equiv x_1^3 \exp\{\cos(x_2^2 + x_4^2)\} + x_1 + 2x_4 + (x_2 + x_1 + x_4)^4) = 0.$$

Then the mapping $F : \mathbb{R}^4 \to \mathbb{R}^2$ is continuously differentiable, as F_1 and F_2 is also so, and that its derivative matrix at the point $0 = (0, 0, 0, 0)$ is

$$F'(0) = \begin{pmatrix} 1 & 0 & 0 & 1 \\ 1 & 0 & 0 & 2 \end{pmatrix}.$$

Thus, the 2×2 matrix

$$\begin{pmatrix} \frac{\partial F_1}{\partial x_1}(0, 0, 0, 0) & \frac{\partial F_1}{\partial x_4}(0, 0, 0, 0) \\ \frac{\partial F_2}{\partial x_1}(0, 0, 0, 0) & \frac{\partial F_2}{\partial x_4}(0, 0, 0, 0) \end{pmatrix} = \begin{pmatrix} 1 & 1 \\ 1 & 2 \end{pmatrix},$$

is invertible. We apply the Implicit Function Theorem to choose a positive number r and continuously differentiable functions $g : B \to \mathbb{R}$ and $h : B \to \mathbb{R}$, where $B = B_r(0, 0)$, such that if $x_2^2 + x_3^2 < r^2$, then $(g(x_2, x_3), x_2, x_3, h(x_2, x_3))$ is a solution of the given system of equations. Moreover, if the point $(x_1, x_2, x_3, x_4) \in \mathbb{R}^4$ is a solution of the given system of equations and if $x_1^2 + x_4^2 < r^2$ and $x_2^2 + x_3^2 < r^2$, then $x_1 = g(x_2, x_3)$ and $x_4 = h(x_2, x_3)$.

Exercises

Exercise 1.25 *Consider the equation*

$$\exp\{x - 2 + (y - 1)^2 - 1\} = 0.$$

Show that Dini's Theorem applies at the solution $(2, 1)$. Explicitly define the function $g : I \to \mathbb{R}$ that has the property that in a neighbourhood of the solution $(2, 1)$, all the solutions are of the form $(x, g(x))$ for $x \in I$ and check that formula (1.21) holds for the derivative $g' : I \to \mathbb{R}$.

Exercise 1.26 *Consider the given system of equations:*

$$(x^2 + y^2 + z^2)^3 - x + z = 0, \ \cos(x^2 + y^4) + e^z - 2 = 0.$$

Use the Implicit Function Theorem to analyze the solutions of the given systems of equations near the solution 0.

Exercise 1.27 *For $e^{x^2} + y^2 + z - 4xy^3 - 1 = 0$, use the Implicit Function Theorem to analyze the solutions of the given systems of equations near the solution 0.*

Chapter 2
Manifold Theory

2.1 Topological Manifold

Curves and surfaces are the fundamental concepts of studying geometry in a 3-dimensional space. The quest for studying these two concepts in a space of higher dimension yields the concept of manifold theory.

A locally **Euclidean space of dimension** n is a topological space such that every point of this space has a neighbourhood homeomorphic to an open subset of $\mathbb{R}^n$.

A topological manifold M of dimension n, denoted by M^n, is a Hausdorff, second countable, locally Euclidean space of dimension n (Fig. 2.1).

Thus, for each $p \in M$, there exists a neighbourhood U of M and a homeomorphism ϕ of U onto an open subset $\phi(U)$ of $\mathbb{R}^n$. The pair (U, ϕ) is called a **chart**. Each such chart (U, ϕ) on M induces a set of n-real-valued functions on U defined by

$$x^i = u^i \circ \phi, \ i = 1, 2, 3, \ldots, n \tag{2.1}$$

where u^i's are defined by (1.1). The functions $(x^1, x^2, \ldots, x^n)$ are called the **coordinate functions** or a **coordinate system** on U and U is called the domain of the coordinate system. The chart (U, ϕ) is sometimes called an n-**coordinate chart**. From (2.1), one obtains

$$x^i(p) = \phi^i(p), \ \text{by (1.1)}.$$

Thus one can write

$$\phi(p) = \left(x^1(p), x^2(p), \ldots, x^n(p)\right). \tag{2.2}$$

Let (V, ψ) be another chart of p of M such that $p \in U \cap V$.

Let $(y^1, y^2, \ldots, y^n)$ be a local coordinate system on V such that

$$y^i = u^i \circ \psi, \ i = 1, 2, 3, \ldots, n \tag{2.3}$$

$$\psi(p) = \left(y^1(p), y^2(p), \ldots, y^n(p)\right). \tag{2.4}$$

© The Author(s), under exclusive license to Springer Nature Singapore Pte Ltd. 2023
M. Majumdar and A. Bhattacharyya, *An Introduction to Smooth Manifolds*,
https://doi.org/10.1007/978-981-99-0565-2_2

Fig. 2.1 Locally Euclidean space of dimension n

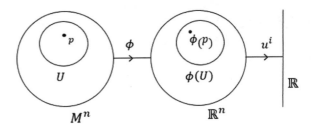

Example 2.1 $\mathbb{R}^n$ is a topological manifold covered by a single chart $(\mathbb{R}^n, I_{\mathbb{R}^n})$ where I is the identity map.

Example 2.2 Every open subset U of $\mathbb{R}^n$ is a topological manifold with chart (U, I_U).

Example 2.3 Every discrete topological space M is a 0-dimensional topological manifold, the charts being given by the pair $(\{p\}, \phi_p)$ where $p \mapsto 0$, $p \in M$.

Example 2.4 Let $f : \mathbb{R}^3 \to \mathbb{R}^3$ be defined by $f(x, y, z) = (x, x^2 + y^2 + z^2 - 1, z)$, $|J| = 2y$. By virtue of Inverse Function Theorem, f is a local diffeomorphism at $p = (x, y, z)$ if and only if $y \neq 0$. Thus, the function f can serve as a local coordinate system at any point p **not** on the x-axis and z-axis.

Remark 2.1 (i) A 1-dimensional manifold is locally homeomorphic to open interval.

(ii) A 2-dimensional manifold is locally homeomorphic to open disc.

Remark 2.2 A topological manifold is

(i) locally connected
(ii) locally compact
(iii) normal and metrizable.

For proof, refer to any standard textbooks on general topology.

Problem 2.1 Does the map $f : \mathbb{R} \to \mathbb{R}$ defined by $f(x) = x^2$ form a chart?

Solution: 1 Note that $f : \mathbb{R} \to \mathbb{R}$ defined by $f(x) = x^2$ is not a homeomorphism on $\mathbb{R}$ (refer to Exercise 1.5). Thus, the given map does not form a chart.

Problem 2.2 Consider the open subsets U and V of the unit circle S^1 of $\mathbb{R}^2$ given by

$$U = \{(\cos\alpha, \sin\alpha) : \alpha \in (0, 2\pi)\}, \quad V = \{(\cos\alpha, \sin\alpha) : \alpha \in (-\pi, \pi)\}$$

and the maps $\phi : U \to \mathbb{R}$ defined by $\phi(\cos\alpha, \sin\alpha) = \alpha$, $\alpha \in (0, 2\pi)$ and $\psi : V \to \mathbb{R}$ defined by $\psi(\cos\alpha, \sin\alpha) = \alpha$, $\alpha \in (-\pi, \pi)$. Prove that (U, ϕ) and (V, ψ) are charts on $\mathbb{R}^2$.

Fig. 2.2 Overlap region

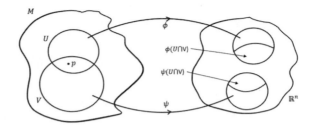

Solution: 2 Note that the two maps ϕ and ψ are homeomorphisms onto the open subsets $(0, 2\pi)$ and $(-\pi, \pi)$ of $\mathbb{R}$, respectively. Consequently, (U, ϕ) and (V, ψ) are charts on $\mathbb{R}^2$.

Problem 2.3 Prove that the graph $y = x^{\frac{2}{3}}$ in $\mathbb{R}^2$ is a topological manifold.

Solution: 3 The graph is a subspace of $\mathbb{R}^2$ and hence it is Hausdorff and second countable. Also $(x, x^{\frac{2}{3}}) \to x$, so it is homeomorphic to $\mathbb{R}$. Consequently, it is locally Euclidean. Hence, the graph $y = x^{\frac{2}{3}}$ in $\mathbb{R}^2$ is a topological manifold.

Problem 2.4 Find the functional relation between the two local coordinate systems defined in the overlap region of a topological manifold.

Solution: 4 Let (U, ϕ) and (V, ψ) be two charts of a point $p \in U \cap V$ of a topological manifold M (Fig. 2.2).

Let $\phi(p) = q \in \mathbb{R}^n$, $q \in \phi(U \cap V)$. Suppose $g : \phi(U \cap V) \to \psi(U \cap V)$ is defined by

$$g(q) = (\psi \circ \phi^{-1})(q). \tag{2.5}$$

Then

$$g(q) = g(\phi(p)) = (\psi \circ \phi^{-1})(\phi(p)), \text{ by (2.5)}$$
$$i.e. \ \ g(q) = \psi(p)$$
$$\text{or} \ \ u^i\big(g(\phi(p))\big) = u^i(\psi(p))$$
$$\text{or} \ \ g^i(\phi(p)) = y^i(p) \text{ by (1.1), (2.3)}$$
$$\text{or} \ \ g^i(x^1(p), x^2(p), \dots, x^n(p)) = y^i(p)$$
$$i.e. \ \ y^i = g^i(x^1, x^2, \dots, x^n).$$

Problem 2.5 Give an example of a non-Hausdorff locally Euclidean space.

Solution: 5 Let $A \subset \mathbb{R}^2$ be such that

$$A = U \bigcup \{(0, 1)\},$$

where $U = \{(x, 0)|x \in \mathbb{R}\}$. Let $U_1 = U \setminus \{(0, 0)\} \cup \{(0, 1)\}$. We define $\phi : U \to \mathbb{R}$ by $\phi(x, 0) = x$ and $\psi : U_1 \to \mathbb{R}$ by

$$\psi(x, 1) = \begin{cases} 0, & x = 0 \\ x, & x \neq 0. \end{cases}$$

Both ϕ and ψ are well-defined on U and U_1, respectively. Also ϕ and ψ are injective maps in $\mathbb{R}$ and $U \bigcup U_1 = A$. So, (U, ϕ) and (U_1, ψ) are charts and hence it is a locally Euclidean space.

Let V_1 be an open neighbourhood of $(0, 0)$ and V_2 be an open neighbourhood of $(0, 1)$ in A. Then both $\phi(U \bigcap V_1)$ and $\psi(U_1 \bigcap V_2)$ are open subsets of $\mathbb{R}$ containing 0. So $\exists\, a \neq 0$ such that $a \in \phi(U \bigcap V_1) \bigcap \psi(U_1 \bigcap V_2)$, which implies $(a, 0) \in V_1 \bigcap V_2$. Hence the topology of A is non-Hausdorff. Thus A fails to form a topological manifold.

Problem 2.6 Consider the cone $C = \{(x, y, z) \in \mathbb{R}^3 : x^2 + y^2 = z^2\}$ with the subspace topology as induced by the usual one of $\mathbb{R}^3$. Prove that C is not a topological manifold.

Solution: 6 Our claim is that the space C is not a locally Euclidean space. It suffices to show that the point $(0, 0, 0) \in C$ does not have a neighbourhood homeomorphic to an open subset of $\mathbb{R}^2$ (Fig. 2.3).

Let U be an open neighbourhood of $(0, 0, 0)$ in C. Let $\phi : U \to V$ be a homeomorphism between U and an open subset V of $\mathbb{R}^2$. Then for some sufficiently small $r > 0$, $\exists$ an open disc $B_r(\phi(0, 0, 0))$ with $\phi(0, 0, 0)$ as its centre such that $B_r(\phi(0, 0, 0)) \subset V$. Now the punctured disc $B_r(\phi(0, 0, 0)) \setminus \{\phi(0, 0, 0)\}$ is connected. But $U \setminus \{(0, 0, 0)\}$ is not connected. In fact,

$$U \setminus \{(0, 0, 0)\} = U_1 \bigcup U_2,$$

where
$$U_1 = \{(x, y, z) \in U : z > 0\}, \quad U_2 = \{(x, y, z) \in U : z < 0\}.$$

So, $U_1 \bigcap U_2 = \phi$ and U_1, U_2 are open in C. Hence, U can be expressed as a disjoint union of two non-empty open subsets of C. Thus C is not a locally Euclidean space.

Problem 2.7 Show that the cross in $\mathbb{R}^2$ with the subspace topology cannot be a topological manifold.

Solution: 7 Our claim is to prove that the cross is not locally Euclidean at the intersection q (Fig. 2.4).

If possible, let us assume that the cross is locally Euclidean of dimension n at the point q. Then $\exists$ a homeomorphism $\phi : V \to B_r(0, 0, 0, \ldots, 0)$, where V is an

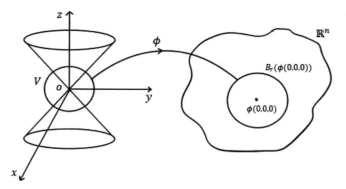

Fig. 2.3 Double cone

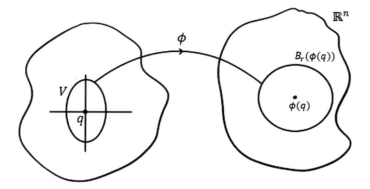

Fig. 2.4 Cross(i)

open neighbourhood of q and $B_r(0, 0, 0, \ldots, 0) \subset \mathbb{R}^n$ is an open ball with centre $(0, 0, 0, \ldots, 0)$ and radius r (sufficiently small enough). Here we assume $\phi(q) = (0, 0, 0, 0, \ldots, 0)$. The homeomorphism $\psi : V \setminus \{q\} \to B_r(0, 0, 0, \ldots, 0) \setminus (0, 0, 0, \ldots, 0)$ acts as a restriction map to a homeomorphism ϕ. Now, if $n \geq 2$ then

$$B_r(0, 0, 0, \ldots, 0) \setminus (0, 0, 0, \ldots, 0)$$

is connected, and if $n = 1$ then it has two connected components (Fig. 2.5).

Since $V \setminus \{q\}$ has four connected components, there $\nexists$ any homeomorphism from $V \setminus \{q\}$ to $B_r(0, 0, 0, \ldots, 0) \setminus \{(0, 0, 0, \ldots, 0)\}$. This contradiction proves that the cross is not locally Euclidean at q.

Fig. 2.5 Cross(ii)

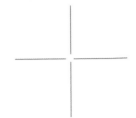

Exercises

Exercise 2.1 *Does the map* $f : \mathbb{R} \to \mathbb{R}$ *defined by* $f(x) = x^{2k}, k \in \mathbb{N},$ *form a chart?*

Exercise 2.2 *Find all points in* $\mathbb{R}^2$, *in a neighbourhood of which the function* $f :$ $\mathbb{R}^2 \to \mathbb{R}^2$ *defined by* $f(x_1, x_2) = (x_1^2 + x_2^2 - 1, x_2)$ *can serve as a local coordinate system.*

Exercise 2.3 *Consider Exercise 1.12 of Sect. 1.1. Can* f *be taken as a local coordinate map?*

Exercise 2.4 *As defined by (2.5),* $h : \psi(U \cap V) \to \phi(U \cap V)$ *is defined by*

$$h(r) = (\phi \circ \psi^{-1})(r), \ r \in \psi(U \cap V). \tag{2.6}$$

Show that the functional relation between the two local coordinate systems defined in the overlap region of a topological manifold is given by

$$x^i = h^i(y^1, y^2, \ldots, y^n).$$

Exercise 2.5 *Let* $X = S^1 \bigcup S^2$ *where*

$$S^1 = \{(x_1, x_2) \in \mathbb{R}^2 : (x_1 - 1)^2 + x_2^2 = 1\}, \ S^2 = \{(x_1, x_2) \in \mathbb{R}^2 : (x_1 + 1)^2 + x_2^2 = 1\}.$$

Suppose X inherits the topology from $\mathbb{R}^2$. *Is X a topological manifold?*

Answers
2.1. Yes.   2.2. Not on the y-axis.   2.3. Yes.   2.5. No.

2.2   Smooth Germs on a Topological Manifold

Let us begin with a definition:

Let M be an n-dimensional topological manifold. Let $f : M \to \mathbb{R}$ be any function. Let $p \in M$. If, for every admissible coordinate chart (U, ϕ) of M satisfying $p \in U, (f \circ \phi^{-1}) : \phi(U) \to \mathbb{R}$ is C^∞ at the point $\phi(p) \in \mathbb{R}^n$, then we say that f is C^∞ at p in M.

Remark 2.3 Note that, if f is C^∞ at p in M, then f is continuous at p.

Theorem 2.1 *Let M be an n-dimensional topological manifold. Let $f : M \to \mathbb{R}$ be any function with $p \in M$. If there exists an admissible coordinate chart (U, ϕ) of M at $p \in U$ such that $(f \circ \phi^{-1}) : \phi(U) \to \mathbb{R}$ is C^∞ at $\phi(p)$, then f is C^∞ at p.*

Proof Let (V, ψ) be any admissible coordinate chart of M with $p \in V$. Our claim is that $(f \circ \psi^{-1}) : \phi(V) \to \mathbb{R}$ is C^∞ at $\psi(p)$. Here

$$f \circ \psi^{-1} = f \circ (\phi^{-1} \circ \phi) \circ \psi^{-1} = (f \circ \phi^{-1}) \circ (\phi \circ \psi^{-1})$$

is C^∞ at $\psi(p)$ as $(f \circ \phi^{-1})$ and $(\phi \circ \psi^{-1})$ are C^∞ at $\phi(p)$ and $\psi(p)$, respectively. This completes the proof.

Let M be an n-dimensional topological manifold. Let $f : M \to \mathbb{R}$ be any function. By f is C^∞ at p in M (or f is smooth on M), we mean that f is C^∞ at every point $p \in M$. The set of all smooth functions $f : M \to \mathbb{R}$ on M is denoted by $C^\infty(M)$.

Remark 2.4 Note that, if f is C^∞ at M, then f is continuous.

Let M and N be respectively n- and m-dimensional topological manifolds. Let $f : M \to N$ be any continuous function. Let $p \in M$. If, for every admissible coordinate chart (U, ϕ) of M satisfying $p \in U$ and (V, ψ) of N with $F(p) \in V$, the mapping

$$\psi \circ (f \circ \phi^{-1}) : \phi(U \cap f^{-1}(V)) \to \psi(V)$$

is C^∞ at $\phi(p)$, then f is C^∞ at p.

Exercise

Exercise 2.6 *Let M and N be respectively n- and m-dimensional topological manifolds. Let $f : M \to N$ be any continuous function. Let $p \in M$. If there exists an admissible coordinate chart (U, ϕ) of M at $p \in U$ and (V, ψ) of N with $f(p) \in V$, the mapping*

$$\psi \circ (f \circ \phi^{-1}) : \phi(U \cap f^{-1}(V)) \to \psi(V)$$

is C^∞ at $\phi(p)$, then f is C^∞ at p.

Let M be an n-dimensional topological manifold, and N be an m-dimensional topological manifold. If there exists a function $f : M \to \mathbb{R}$ such that f is a diffeomorphism from M onto N, then we say that the manifolds M and N are isomorphic (or diffeomorphic).

Let M be an n-dimensional topological manifold. Let $p \in M$. By $C_p^\infty(M)$ (or simply C_p^∞), we mean the collection of all real-valued functions f whose $Dom\, f$ is an open neighbourhood of $p \in M$, and for every admissible coordinate chart U, ϕ of M satisfying $p \in U$,

$$(f \circ \phi^{-1}) : \phi(Dom\, f \cap U) \to \mathbb{R}$$

is C^∞ at the point $\phi(p) \in \mathbb{R}^n$. Observe that if $f \in C_p^\infty(M)$, then f is continuous on some open neighbourhood of p.

Let M be an n-dimensional topological manifold. For every $f, g \in C^\infty(M)$, we define $(f + g) : M \to \mathbb{R}$ as follows: for every $x \in M$,

$$(f + g)(x) = f(x) + g(x).$$

For every $f \in C^\infty(M)$, and for every real t, we define $tf : M \to \mathbb{R}$ as follows: for every $x \in M$,

$$(tf)(x) = tf(x).$$

Remark 2.5 It is clear that the set $C^\infty(M)$, together with vector addition, and scalar multiplication defined as above, constitutes a real linear space.

For every $f, g \in C^\infty(M)$, we define $(f \cdot g) : M \to \mathbb{R}$ as follows: for every $x \in M$,

$$(f \cdot g)(x) = f(x)g(x).$$

Remark 2.6 It is easy to see that $C^\infty(M)$ is an algebra.

Let M be an n-dimensional topological manifold. Let $p \in M$. Let γ_1 and γ_2 be, in $\Gamma_p(M)$, set of all parametrized curves in M through p. Then the relation '$\prec$'on $\Gamma_p(M)$ is defined as follows: by $\gamma_1 \prec \gamma_2$, we mean that for any admissible coordinate chart (U, ϕ) of M at $p \in U$,

$$(\phi \circ \gamma_1)'(0) = (\phi \circ \gamma_2)'(0).$$

Proposition 2.1 *Let M be an n-dimensional topological manifold. Let $p \in M$. Let γ_1 and γ_2 be in $\Gamma_p(M)$. If there exists an admissible coordinate chart (U, ϕ) of M at $p \in U$ such that $(\phi \circ \gamma_1)'(0) = (\phi \circ \gamma_2)'(0)$, then $\gamma_1 \prec \gamma_2$.*

Proof Let us take any admissible coordinate chart (ψ, V) of M satisfying $p \in V$. Our claim is $(\psi \circ \gamma_1)'(0) = (\psi \circ \gamma_2)'(0)$. Now

$$(\psi \circ \gamma_1)'(0) = ((\psi \circ \phi^{-1}) \circ (\phi \circ \gamma_1))'(0)$$
$$= ((\psi \circ \phi^{-1})'(\phi \circ \gamma_1)(0))((\phi \circ \gamma_1)'(0))$$
$$= ((\psi \circ \phi^{-1})'(\phi \circ \gamma_1)(0))((\phi \circ \gamma_2)'(0))$$
$$= ((\psi \circ \phi^{-1})'(\phi(p))((\phi \circ \gamma_2)'(0)), \text{ where } \gamma_1(0) = p$$
$$= ((\psi \circ \phi^{-1})'(\phi \circ \gamma_2)(0))((\phi \circ \gamma_2)'(0))$$
$$= ((\psi \circ \phi^{-1}\circ) \circ (\phi \circ \gamma_2))'$$
$$= (\psi \circ (\phi^{-1} \circ \phi) \circ \gamma_2)'(0)$$
$$= (\psi \circ \gamma_2)'(0).$$

Problem 2.8 Let M be an n-dimensional topological manifold. Let $p \in M$. Then, the relation '$\prec$' on $\Gamma_p(M)$ is an equivalence relation.

Solution: 8 • Reflexive: Since M be an n-dimensional topological manifold and $p \in M$, there exists an admissible coordinate chart (U, ϕ) of M such that $p \in U$. Since $(\phi \circ \gamma)'(0) = (\phi \circ \gamma)'(0)$, then $\gamma \prec \gamma$.
• Symmetric: Let $\gamma_1 \prec \gamma_2$ hold. Let us take an admissible coordinate chart (U, ϕ) of M such that $p \in U$. Since $\gamma_1 \prec \gamma_2$, then $(\phi \circ \gamma_1)'(0) = (\phi \circ \gamma_2)'(0)$ implies $\gamma_2 \prec \gamma_1$.
• Transitive: Let $\gamma_1 \prec \gamma_2$ and $\gamma_2 \prec \gamma_3$ holds. We are to prove $\gamma_1 \prec \gamma_3$. Let us take an admissible coordinate chart (U, ϕ) of M such that $p \in U$. Since $\gamma_1 \prec \gamma_2$, therefore $(\phi \circ \gamma_1)'(0) = (\phi \circ \gamma_2)'(0)$. Also for $\gamma_2 \prec \gamma_3$, we have $(\phi \circ \gamma_2)'(0) = (\phi \circ \gamma_3)'(0)$. Combining the last two relations, we obtain $(\phi \circ \gamma_1)'(0) = (\phi \circ \gamma_3)'(0)$ implies $\gamma_1 \prec \gamma_3$.

Remark 2.7 Let M be an n-dimensional topological manifold. Let $p \in M$. By the last proposition, the quotient set $\Gamma_p(M)/ \prec$ is the collection of all equivalence classes $[\gamma]$, where $\gamma \in \Gamma_p(M)$. Thus,

$$\Gamma_p(M)/ \prec = \{[\gamma] : \gamma \in \Gamma_p(M)\}$$

where

$$[\gamma] = \{\gamma_i \in \Gamma_p(M) : \gamma \prec \gamma_i\}.$$

Let M be an n-dimensional topological manifold. Let $p \in M$. Let us define a function $\phi_* : \Gamma_p(M)/ \prec \to \mathbb{R}^n$ by

$$\phi_*([\gamma]) = (\phi \circ \gamma)'(0).$$

Here ϕ_* is well-defined.

Problem 2.9 Let M be an n-dimensional topological manifold. Let $p \in M$. Suppose (U, ϕ) is any admissible coordinate chart of M such that $p \in U$. Then ϕ_* as defined above is bijective.

Solution: 9 Injectivity: Let $[\gamma_1] \in \Gamma_p(M)/ \prec$ and $[\gamma_2] \in \Gamma_p(M)/ \prec$, where $\gamma_1, \gamma_2 \in \Gamma_p(M)$. Suppose $\phi_*([\gamma_1]) = \phi_*([\gamma_2])$. We are to prove $[\gamma_1] = [\gamma_2]$, i.e. $\gamma_1 \prec \gamma_2$. Now

$$\phi_*([\gamma_1]) = \phi_*([\gamma_2]) \Rightarrow (\phi \circ \gamma_1)'(0) = (\phi \circ \gamma_2)'(0) \Rightarrow \gamma_1 \prec \gamma_2, \text{ by Proposition 2.1.}$$

Surjectivity: Let $q \in \mathbb{R}^n$. We are to find $\gamma \in \Gamma_p(M)$ such that

$$\phi_*([\gamma]) = (\phi \circ \gamma)'(0) = \lim_{t \to 0} \frac{(\phi \circ \gamma)(t) - (\phi \circ \gamma)(0)}{t} = \lim_{t \to 0} \frac{\phi(\gamma(t)) - \phi(\gamma(0))}{t} = \lim_{t \to 0} \frac{\phi(\gamma(t)) - \phi(p)}{t} = q.$$

Let us define a function $\gamma_1 : (-1, 1) \to \mathbb{R}^n$ by $\gamma_1(t) = tq + \phi(p)$. Set

$$\gamma = \phi^{-1} \circ \gamma_1.$$

Then $\gamma(0) = p$ holds. Here γ_1 is continuous on $(-1, 1)$. Since (U, ϕ) is a coordinate chart of M, ϕ^{-1} is $1 - 1$, onto and continuous at $\phi(U)$, which is an open subset of $\mathbb{R}^n$. Since $p \in U$, $\phi(p) \in \phi(U)$. Also ϕ^{-1} is continuous at $\gamma_1(0) = \phi(p) \in \phi(U)$. Moreover, $\phi(U)$ forms an open neighbourhood of $\gamma_1(0)$. Since γ_1 is continuous, $\phi(U)$ being an open neighbourhood of $\gamma_1(0) \, \exists \, \delta > 0$ with $\delta < 1$ and for every $t \in (-\delta, \delta)$, we have $\gamma_1(t) \in \phi(U)$. Hence, $\gamma(t) = (\phi^{-1} \circ \gamma_1)(t) = \phi^{-1}(\gamma_1(t)) \in U$. Since $\gamma(t) \in U$ for every $t \in (-\delta, \delta)$, it implies that γ is defined on $(-\delta, \delta)$ for some $\delta > 0$. Now for every $t \in (-\delta, \delta)$,

$$\gamma(t) = (\phi^{-1} \circ \gamma_1)(t) = \phi^{-1}(tq + \phi(p)) \in U,$$

this shows that γ maps from $(-\delta, \delta)$ to U. Furthermore, γ_1 is C^∞ at every $t \in (-\delta, \delta)$, $\phi \circ \gamma$ is C^∞ at every t and (U, ϕ) being an admissible coordinate chart of M such that $\gamma(t) \in U$, by virtue of Exercise 2.6, γ is C^∞ at every $t \in (-\delta, \delta)$. Thus $\gamma \in \Gamma_p(M)$.
Finally,

$$\lim_{t \to 0} \frac{\phi(\gamma(t)) - \phi(p)}{t} = \lim_{t \to 0} \frac{\phi(\phi^{-1} \circ \gamma_1(t)) - \phi(p)}{t} = \lim_{t \to 0} \frac{\phi(\gamma_1(t)) - \phi(p)}{t} = \lim_{t \to 0} \frac{\phi(tq + \phi(p)) - \phi(p)}{t} = q.$$

This completes the proof.

Since ϕ_* is $1 - 1$ and onto, ϕ_*^{-1} exists and is also $1 - 1$ and onto. Let us define a binary composition $\oplus$ and external composition $\odot$ on $\Gamma_p(M)/ \prec$ as follows.
For every $[\gamma_1], [\gamma_2] \in \Gamma_p(M)/ \prec$, where $\gamma_1, \gamma_2 \in \Gamma_p(M)/ \prec$,

$$[\gamma_1] \oplus [\gamma_2] = \phi_*^{-1}(\phi_*([\gamma_1]) + \phi_*([\gamma_2])),$$

and

$$t \odot [\gamma_1] = \phi_*^{-1}(t(\phi_*(\gamma_1))), \; \forall \, t \in \mathbb{R}.$$

Fig. 2.6 Germ

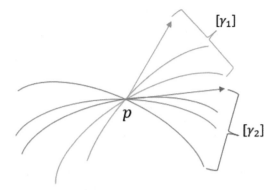

$[\gamma_1]$

p

$[\gamma_2]$

Remark 2.8 The quotient set $(\Gamma_p(M)/\prec, \oplus, \odot)$ forms a real linear space.

Remark 2.9 Since $\Gamma_p(M)/\prec$ is a real linear space, ϕ_* is $1-1$ and onto, ϕ_* is linear, i.e.

$$\phi_*([\gamma_1] \oplus (t \odot [\gamma_2])) = \phi_*([\gamma_1]) + t\,\phi_*([\gamma_2]),$$

ϕ_* is an isomorphism between $\Gamma_p(M)/\prec$ and $\mathbb{R}^n$. Hence the dimension of $(\Gamma_p(M)/\prec$ is n. The set $\Gamma_p(M)/\prec$ is denoted by $T_p(M)$ (Fig. 2.6).

Problem 2.10 Let M be an n-dimensional topological manifold and $p \in M$. Let $\preceq$ be a relation on $C_p^\infty(M)$ defined as follows: for every $f, g \in C_p^\infty(M)$, by $f \preceq g$ we mean that there exists an open neighbourhood V of p such that $f(x) = g(x)$ for every $x \in V$. Then $\preceq$ is an equivalence relation over $C_p^\infty(M)$.

Solution: 10 Reflexivity: Left to the reader.
   Surjectivity: Left to the reader.
   Transitivity: Let $f \preceq g$ and $g \preceq h$ hold where $f, g, h \in C_p^\infty(M)$. Since $f \preceq g$, from the definition of $\preceq$, there exists an open neighbourhood V of p such that $f(x) = g(x)$ for every $x \in V$. By similar reason, for $g \preceq h$, there exists an open neighbourhood W of p such that $g(x) = h(x)$ for every $x \in W$. Since V, W are open neighbourhoods of $p \in M$, therefore $V \cap W$ is an open neighbourhood of $p \in M$. Also $f(x) = h(x)$ for every $x \in V \cap W$. Hence $f \preceq h$ holds.

Remark 2.10 The quotient set $C_p^\infty(M)/\preceq$, of all equivalence classes, is denoted by $\mathfrak{F}_p(M)$. Thus

$$\mathfrak{F}_p(M) = \{[f] : f \in C_p^\infty(M)\},$$

where

$$[f] = \{g : g \in C_p^\infty(M), g \preceq f\}$$

is a C^∞-germ at p on M. So, the members of $\mathfrak{F}_p(M)$ are called C^∞-germs at p on M.

Let M be an n-dimensional topological manifold and $p \in M$. For every $f, g \in C_p^\infty(M)$, we define $(f + g) : Dom\ f \cap Dom\ g \to \mathbb{R}$ as follows: for every $x \in Dom\ f \cap Dom\ g$,

$$(f + g)(x) = f(x) + g(x).$$

For every $f, g \in C_p^\infty(M)$, we define $f \cdot g : Dom\ f \cap Dom\ g \to \mathbb{R}$ as follows: for every $x \in Dom\ f \cap Dom\ g$,

$$(f \cdot g)(x) = f(x)g(x).$$

Remark 2.11 Let M be an n-dimensional topological manifold and $p \in M$. Let $f, g \in C_p^\infty(M)$. Then $f + g, f \cdot g : Dom\ f \cap Dom\ g \to \mathbb{R}$, is in $C_p^\infty(M)$. Also for any $t \in \mathbb{R}$,
$tf : Dom\ f \to \mathbb{R}$ is in $C_p^\infty(M)$.

Let M be an n-dimensional topological manifold and $p \in M$. For every $f, g \in C_p^\infty(M)$ and $t \in \mathbb{R}$, we define

$$[f] + [g] = [f + g],\ t[f] = [tf],\ [f][g] = [f \cdot g].$$

Remark 2.12 $\mathfrak{F}_p(M)$ forms a real linear space. Also, $\mathfrak{F}_p(M)$ forms an algebra.

Remark 2.13 Let M be an n-dimensional topological manifold and $p \in M$. If $f, g \in C_p^\infty(M)$, then $(f + g) \in C_p^\infty(M)$.

Remark 2.14 Let M be an n-dimensional topological manifold and $p \in M$. If $f, g \in C_p^\infty(M)$, then $(f \cdot g) \in C_p^\infty(M)$.

Remark 2.15 Let M be an n-dimensional topological manifold and $p \in M$. If $f \in C_p^\infty(M)$ and $t \in \mathbb{R}$, then $tf \in C_p^\infty(M)$.

Problem 2.11 Let M be an n-dimensional topological manifold and $p \in M$. Let $\gamma \in \Gamma_p(M)$. Let $[f] \in \mathfrak{F}_p(M)$, where $f \in C_p^\infty(M)$. Then

$$\lim_{t \to 0} \frac{(f \circ \gamma)(0 + t) - (f \circ \gamma)(0)}{t}$$

exists.

Solution: 11 Since $\gamma \in \Gamma_p(M)$, it follows from the definition of $\Gamma_p(M)$ that $\exists\ \delta > 0$ such that $\gamma : (-\delta, \delta) \to M$, $\gamma(0) = p$ and γ is a smooth map on $(-\delta, \delta)$, and so is continuous on $(-\delta, \delta)$. Here, the function $f : Dom\ f \to \mathbb{R}$ is in $\in C_p^\infty(M)$, where $Dom\ f$ is an open neighbourhood of $p \in M$, so f is continuous on some open neighbourhood $U(\subseteq Dom\ f)$ of p. Since $\gamma(0) = p$ and $p \in Dom\ f$, therefore $0 \in Dom\ (f \circ \gamma)$. Moreover, as γ is continuous with $\gamma(0) = p$ and U being an open neighbourhood of $p, \exists \epsilon > 0$ such that $\epsilon < \delta$ and for every $t \in (-\epsilon, \epsilon)$, we have $\gamma(t) \in U(\subseteq Dom\ f)$, and hence $(f \circ \gamma)(t) = f(\gamma(t)) \in \mathbb{R}$. This implies that $(-\epsilon, \epsilon) \subseteq$

$Dom\ (f \circ \gamma)$. Hence 0 is an interior point of $f \circ \gamma$. Here, $\phi \circ \gamma : (-\delta, \delta) \to \mathbb{R}^n$ is C^∞ at the point 0 in $\mathbb{R}$ and $(f \circ \phi^{-1}) : \phi(Dom\ f \cap U) \to \mathbb{R}$ is C^∞ at the point $(\phi \circ \gamma)(0)$ in $\mathbb{R}^n$. So the composite function $f \circ \gamma$ is C^∞ at the point 0 in $\mathbb{R}$. Hence, $\dfrac{d(f \circ \gamma)(t)}{dt}\Big|_{t=0}$, i.e. $\lim\limits_{t \to 0} \dfrac{(f \circ \gamma)(0 + t) - (f \circ \gamma)(0)}{t}$ exists.

So the following definition is well defined:

Let M be an n-dimensional topological manifold and $p \in M$. Let $\gamma \in \Gamma_p(M)$. Let $[f] \in \mathfrak{F}_p(M)$, where $f \in C_p^\infty(M)$. Then

$$\lim_{t \to 0} \frac{(f \circ \gamma)(0 + t) - (f \circ \gamma)(0)}{t}$$

exists, and is denoted by $< \gamma, [f] >$. Thus

$$< \gamma, [f] > \equiv \lim_{t \to 0} \frac{(f \circ \gamma)(0 + t) - (f \circ \gamma)(t)}{t},$$

i.e.

$$< \gamma, [f] > \equiv \frac{d(f \circ \gamma)(t)}{dt}\Big|_{t=0}.$$

Theorem 2.2 *Let M be an n-dimensional topological manifold and $p \in M$. Let $\gamma \in \Gamma_p(M)$. Suppose $[f], [g] \in \mathfrak{F}_p(M)$, where $f, g \in C_p^\infty(M)$. Then*

1. $< \gamma, [f] + [g] > = < \gamma, [f] > + < \gamma, [g] >$
2. $< \gamma, c[f] > = c < \gamma, [f] >$, *where $c \in \mathbb{R}$.*

In other words, $<, >$ is linear in the second variable.

Proof 1. Here,

$$< \gamma, [f] + [g] > = <\gamma, [f + g] > = \lim_{t \to 0} \frac{((f + g) \circ \gamma)(0 + t) - ((f + g) \circ \gamma)(0)}{t}$$

$$= \lim_{t \to 0} \frac{(f + g)(\gamma)(t) - (f + g)(\gamma)(0)}{t}$$

$$= \lim_{t \to 0} \frac{\{(f(\gamma)(t)) - (f(\gamma)(0))\} + \{(g(\gamma)(t)) - (g(\gamma)(0))\}}{t}$$

$$= \lim_{t \to 0} \frac{\{(f(\gamma)(t)) - (f(\gamma)(0))\}}{t} + \frac{\{(g(\gamma)(t)) - (g(\gamma)(0))\}}{t}$$

$$= \lim_{t \to 0} \frac{\{(f(\gamma)(0 + t)) - (f(\gamma)(0))\}}{t} + \frac{\{(g(\gamma)(0 + t)) - (g(\gamma)(0))\}}{t}$$

$$= <\gamma, [f] > + < \gamma, [g] > .$$

2. Left to the reader.

Theorem 2.3 *Let M be an n-dimensional topological manifold and $p \in M$. Let $\gamma \in \Gamma_p(M)$. Suppose $[f], [g] \in \mathfrak{F}_p(M)$, where $f, g \in C_p^\infty(M)$. Further, suppose $[\gamma] \in T_p(M)$, where $\gamma \in \Gamma_p(M)$. Then*

1. $[\gamma]([f] + [g]) = [\gamma]([f]) + [\gamma]([g])$.
2. $[\gamma]([tf]) = t([\gamma]([f])$, *where* $t \in \mathbb{R}$.
3. $[\gamma]([f][g]) = ([\gamma][f])(g(p)) + f(p)([\gamma][g])$.

Proof   1. Left to the reader.
2. Left to the reader.
3. Here

$$[\gamma]([f][g]) = [\gamma]([f \cdot g]) = \frac{d((f \cdot g) \circ \gamma)(t)}{dt}\Big|_{t=0} = \frac{d((f \cdot g)(\gamma(t)))}{dt}\Big|_{t=0}$$

$$= \frac{d(f(\gamma(t)) \cdot g(\gamma(t)))}{dt}\Big|_{t=0}$$

$$= \frac{d((f \circ \gamma)(t) \cdot (g \circ \gamma)(t))}{dt}\Big|_{t=0}$$

$$= \frac{d(f \circ \gamma)(t)}{dt}\Big|_{t=0}(g \circ \gamma)(0) + (f \circ \gamma)(0)\frac{d(g \circ \gamma)(t)}{dt}\Big|_{t=0}$$

$$= ([\gamma][f])g(\gamma(0)) + f(\gamma(0))([\gamma][g])$$

$$= ([\gamma][f])(g(p)) + f(p)([\gamma][g]).$$

2.3   Smooth Manifold

We are now going to introduce a differentiable structure on a topological manifold. For this, let us at first introduce **compatible charts** or C^∞**-related charts**.

Two charts (U, ϕ) and (V, ψ) on a topological manifold M are said to be C^∞**-compatible** or C^∞**-related** if

$$\begin{cases} \text{either } U \cap V = \phi, & \text{or;} \\ U \cap V \neq \phi, & \text{and the} \\ \text{transition maps } \phi \circ \psi^{-1} : \psi(U \cap V) \to \phi(U \cap V), \text{ and} \\ \psi \circ \phi^{-1} : \phi(U \cap V) \to \psi(U \cap V) \text{ are of class } C^\infty. \end{cases} \quad (2.7)$$

In short, we say **compatible**. These two maps are called the transition functions between the charts. If $U \cap V = \phi$, then the two charts are obviously C^∞-compatible.

Problem 2.12 Prove that the compatibility of charts is not an equivalence relation.

Solution: 12 Let (U, ϕ), (V, ψ) and (W, φ) be three charts on a topological manifold.

(i) Here, (U, ϕ) is C^∞-compatible to itself as $\phi \circ \phi^{-1} = I$ is C^∞. Hence compatibility of charts is reflexive.

(ii) Let us assume that the chart (U, ϕ) is C^∞-compatible to (V, ψ). Then by (2.7), we have

$$(*) \begin{cases} \phi \circ \psi^{-1} : \psi(U \cap V) \to \phi(U \cap V) \text{ and} \\ \psi \circ \phi^{-1} : \phi(U \cap V) \to \psi(U \cap V) \end{cases}$$

are of class C^∞, where $p \in U \cap V$. To prove (V, ψ) is C^∞-compatible to (U, ϕ), we have to show

$$\begin{cases} \psi \circ \phi^{-1} : \phi(U \cap V) \to \psi(U \cap V) \text{ and} \\ \phi \circ \psi^{-1} : \psi(U \cap V) \to \phi(U \cap V) \end{cases}$$

are of class C^∞, which follows from $(*)$. Consequently, the compatibility of charts is symmetric.

(iii) Let us assume that the chart (U, ϕ) is C^∞-compatible to (V, ψ) and that the chart (V, ψ) is C^∞-compatible to (W, φ). To prove the compatibility of charts is transitive, we need to show (U, ϕ) is C^∞-compatible to (W, φ).

Since (V, ψ) is C^∞-compatible to (W, φ), therefore

$$(**) \begin{cases} \psi \circ \varphi^{-1} : \varphi(V \cap W) \to \phi(V \cap W) \text{ and} \\ \varphi \circ \psi^{-1} : \psi(V \cap W) \to \varphi(V \cap W) \end{cases}$$

are of class C^∞.
Note that

$$\phi \circ \varphi^{-1} = (\phi \circ \psi^{-1}) \circ (\psi \circ \varphi^{-1}) \quad \text{and} \quad \varphi \circ \phi^{-1} = (\varphi \circ \psi^{-1}) \circ (\psi \circ \phi^{-1}).$$

Now for any $p \in U \cap V \cap W$, we have $\phi \circ \varphi^{-1}$ and $\varphi \circ \phi^{-1}$ are C^∞ on $\varphi(U \cap V \cap W)$ and $\phi(U \cap V \cap W)$, respectively. But, in particular, if we consider any $p \in (U \cap W) \setminus (U \cap V \cap W)$ then $\phi \circ \varphi^{-1}$ and $\varphi \circ \phi^{-1}$ fails to be C^∞ on $\varphi(U \cap W)$ and $\phi(U \cap W)$, respectively. Hence compatibility of charts is not transitive.

This proves C^∞-compatibility of charts is not an equivalence relation (Fig. 2.7).

An **atlas** on a topological manifold M is a collection of pairwise C^∞-compatible charts $\{(U_\alpha, \phi_\alpha) : \alpha \in I\}$ such that

$$\begin{cases} (i) \bigcup_{\alpha \in I} U_\alpha = M \\ (ii) \, \phi_\alpha \circ \phi_\beta^{-1} \text{ or } \phi_\beta \circ \phi_\alpha^{-1} \text{ is } C^\infty \text{ on } \phi_\beta(U_\alpha \cap U_\beta) \text{ or on } \phi_\alpha(U_\alpha \cap U_\beta). \end{cases}$$
$$(2.8)$$

A **differential structure** on M is an atlas $\mathbb{A}$ which is maximal, i.e. if (U, ϕ) is another chart such that $\phi \circ \phi_\alpha^{-1}$ and $\phi_\alpha \circ \phi^{-1}$ respectively on $\phi_\alpha(U \cap U_\alpha)$ and $\phi(U \cap U_\alpha)$ are of class C^∞, then $(U, \phi) \in \mathbb{A}$.

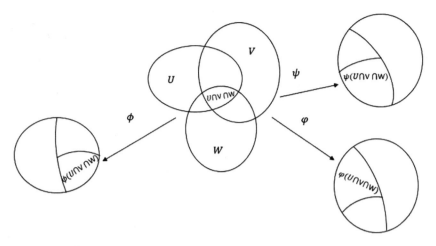

Fig. 2.7 $\varphi \circ \phi^{-1}$ is C^{∞} on $\phi(U \cap V \cap W)$

A **smooth manifold** or a **differentiable manifold of class** C^{∞} of dimension n is a pair $(M, \mathbb{A})$ where M is a topological manifold of dimension n and $\mathbb{A}$ is an atlas.

Proposition 2.2 *Let $\mathbb{A}$ be an atlas on a smooth manifold M. Then, there is a unique differential structure Ω on M such that $A \subset \Omega$.*

Proof Let Ω be the set of all charts on M which are C^{∞}-related to every chart of $\mathbb{A}$. Then clearly, $A \subset \Omega$.

If Ω is an atlas, by definition, it is maximal and has a differential structure. We are now going to prove the uniqueness of such Ω.

If possible, let Ω' be another differential structure on M with the desired property. Then $\mathbb{A} \subset \Omega' \subset \Omega$, by definition of Ω. But by the maximality of Ω', $\Omega' = \Omega$.

Now, let (U, ϕ) and (V, ψ) be any two charts of Ω. We wish to show that $\phi \circ \psi^{-1} : \psi(U \cap V) \to \phi(U \cap V)$ is of class C^{∞}. For this, let $x \in \psi(U \cap V)$ and $(U_i, \phi_i) \in \Omega$ be such that $\psi^{-1}(x) \in U_i$. Now

$$\phi \circ \phi_i^{-1} : \phi_i(U \cap U_i) \to \mathbb{R}^n \text{ and}$$

$$\phi_i \circ \psi^{-1} : \psi(U_i \cap V) \to \mathbb{R}^n$$

are of class C^{∞}.

Let W be an open neighbourhood of x such that $W \subset \phi(U \cap V \cap U_i)$. Now on W, $\phi \circ \psi^{-1} = \phi \circ \phi_i^{-1} \circ \phi_i \circ \psi^{-1}$ is of class C^{∞}, as a composition of two C^{∞}-functions is also so. This completes the proof.

Remark 2.16 Now, it must be clear that to introduce differential structure, one needs to find a chart or a coordinate map. We proceed as follows.

Let $f^i : i = 1, 2, 3, \ldots, n$ be n-real-valued C^{∞} functions defined on M. Let the set $\{f^i\}$ be non-vanishing Jacobian at $p \in M$. Then by Inverse Function Theorem, there exists a neighbourhood V of p and a neighbourhood U of $\left(f^i(p) : i =\right.$

$1, 2, 3, \ldots, n)$ such that $f = (f^1, f^2, \ldots, f^n)$ mapping V into U in $1-1$ manner has an C^∞-inverse. Let it be denoted by ϕ^{-1}. Then (V, ϕ) is the desired chart.

Consider $f : \mathbb{R}^2 \to \mathbb{R}^2$ defined by $f(x, y) = (x^2 + 3y^2, xy)$. Thus

$$f^1(x, y) = x^2 + 3y^2, \quad f^2(x, y) = xy.$$

Therefore

$$|J| = \begin{vmatrix} 2x & 6y \\ y & x \end{vmatrix} = 2x^2 - 6y^2 \neq 0, \text{ if and only if } x \pm \sqrt{3}y \neq 0.$$

Consequently, for all other points of $\mathbb{R}^2$, we can find a neighbourhood V of $p = (x, y)$ and the neighbourhood U of (f^1, f^2) which is mapped by ϕ^{-1} to V. Thus (V, ϕ) is a chart.

Example 2.5 $\mathbb{R}^n$ is a smooth manifold with respect to the atlas $\{(U, \phi)\}$ where $U = \mathbb{R}^n$ and ϕ is the identity map.

Example 2.6 Any open subset W of a smooth manifold M^n is a smooth manifold of the same dimension. For, if $\{(U_\alpha, \phi_\alpha)\}$ is an atlas of M, then $\left\{ U_\alpha \cap W, \ \phi_\alpha|_{U_\alpha \cap W} \right\}$ is an atlas for W where

$$\phi_\alpha|_{U_\alpha \cap W} : U_\alpha \cap W \to \mathbb{R}^n$$

is of class C^∞.

Remark 2.17 Any chart is said to be compatible with an atlas $\mathcal{A}$, if it is compatible with all the charts of $\mathcal{A}$.

Problem 2.13 Let $\mathcal{A}$ be an atlas on a topological manifold. If two charts (V, ψ) and (W, θ) are both compatible with $\mathcal{A}$, then they are compatible with each other.

Solution: 13 Let $p \in V \cap W$. By (2.7), let $\mathcal{A} = \{(U_\alpha, \phi_\alpha) : \alpha \in \wedge\}$ be the atlas. Since (V, ψ) and (W, θ) are compatible with $\mathcal{A}$, $p \in U_\alpha$ for some α, i.e. $p \in V \cap W \cap U_\alpha$. Now

$$\theta \circ \psi^{-1} = (\theta \circ \phi_\alpha^{-1}) \circ (\phi_\alpha \circ \psi^{-1}).$$

Again $\phi_\alpha \circ \psi^{-1} : \psi(V \cap W \cap U_\alpha) \to \phi_\alpha(V \cap W \cap U_\alpha)$ is C^∞, as (V, ψ) is compatible with $\mathcal{A}$. Finally,

$$\phi_\alpha^{-1} \circ (\phi_\alpha \circ \psi^{-1}) : \psi(V \cap W \cap U_\alpha) \to V \cap W \cap U_\alpha \text{ is } C^\infty \text{ i.e.}$$
$$(\theta \circ \phi_\alpha^{-1}) \circ (\phi_\alpha \circ \psi^{-1}) : \psi(V \cap W \cap U_\alpha) \to \theta(V \cap W \cap U_\alpha) \text{ is } C^\infty$$

on $\psi(V \cap W \cap U_\alpha)$ and hence on $\psi(p)$. Consequently, $\theta \circ \psi^{-1}$ is C^∞. Similarly, it can be shown that $\psi \circ \theta^{-1}$ is C^∞ on $\theta(V \cap W \cap U_\alpha)$ and hence on $\theta(p)$. Thus, (V, ψ) and (W, θ) are both compatible with each other.

Problem 2.14 Prove that $\mathcal{A} = \{(U, \phi), (V, \psi)\}$, where U, V, ϕ, ψ are defined in Problem 2.2 is an atlas on S^1.

Solution: 14 In Problem 2.2, it has been proved that (U, ϕ) and (V, ψ) are charts on $\mathbb{R}^2$, where

(i) $S^1 = U \cup V$

(ii) Now

$$\phi^{-1} : \phi(U \cap V) \to U \cap V \text{ and } \psi \circ \phi^{-1} : \phi(U \cap V) \to \psi(U \cap V)$$

such that

$$(\psi \circ \phi^{-1})(\alpha) = \psi(\cos \alpha, \sin \alpha) = \begin{cases} \alpha, & \text{if } \alpha \in (0, \pi) \\ \alpha - 2\pi, & \text{if } \alpha \in (\pi, 2\pi) \end{cases}$$

and hence C^∞. Also

$$\phi \circ \psi^{-1} : \psi(U \cap V) \to \phi(U \cap V)$$

is such that $\phi(U \cap V)(\alpha) = \phi(\cos \alpha, \sin \alpha) = \alpha, \alpha \in (0, \pi)$, and hence C^∞. Thus $\mathcal{A} = \{(U, \phi), (V, \psi)\}$ is an atlas on $\mathbb{R}^2$.

Problem 2.15 Prove that S^1 is a 1-dimensional manifold.

Solution: 15 Let $S^1 = \{(x, y) : (x, y) \in \mathbb{R}^2, \sqrt{x^2 + y^2} = 1\}$ be a unit circle in $\mathbb{R}^2$. We give S^1, the topology of a subspace of $\mathbb{R}^2$. Let

$$U_1 = \{(x, y) \in S^1 : y > 0\}, \qquad U_2 = \{(x, y) \in S^1 : y < 0\}$$
$$U_3 = \{(x, y) \in S^1 : x > 0\}, \qquad U_4 = \{(x, y) \in S^1 : x < 0\}.$$

Then each U_i is an open subset of S^1 and $S^1 = \bigcup U_i, i = 1, 2, 3, 4$. We define

$$\phi_1 : U_1 \to \mathbb{R} \text{ be such that } \phi_1(x, y) = x$$
$$\phi_2 : U_2 \to \mathbb{R} \text{ be such that } \phi_2(x, y) = x$$
$$\phi_3 : U_3 \to \mathbb{R} \text{ be such that } \phi_3(x, y) = y$$
$$\phi_4 : U_1 \to \mathbb{R} \text{ be such that } \phi_4(x, y) = y.$$

Then each ϕ_i is a homeomorphism on $\mathbb{R}$ and hence each (U_i, ϕ_i) is a chart of S^1. Now $U_1 \cap U_2 = \Phi$, so $(U_i, \phi_i), i = 1, 2$ are C^∞-related. Further, $U_1 \cap U_3 \neq \Phi$. Let $p \in U_1 \cap U_3$. Then $(\phi_1 \circ \phi_3^{-1})(y) = x$, $(\phi_3 \circ \phi_1^{-1})(x) = y$ are of class C^∞. Proceeding in this manner, it can be shown that $\mathbb{A} = \{(U_i, \phi_i) : i = 1, 2, 3, 4\}$ is an atlas of S^1 and hence $(S^1, \mathbb{A})$ is a 1-dimensional manifold.

Problem 2.16 Prove that the topological space $\mathbb{M}(m \times n, \mathbb{R})$ of all $m \times n$ order matrices with real entries form a smooth manifold of dimension mn.

Solution: 16 Let us define the map $\phi : \mathbb{M} \to \mathbb{R}^{mn}$ by

$$\phi(A) = (a_{11}, a_{12}, \dots, a_{1n}; \dots, \dots; a_{m1}, a_{m2}, \dots, a_{mn}),$$

where $A = (a_{ij})$, $i = 1, 2, \dots, m$; $j = 1, 2, \dots, n$. Here ϕ is one-to-one and onto. Moreover with the topology induced by $\mathbb{R}^{mn}$ on $\mathbb{M}$, ϕ is a homeomorphism. So $(\mathbb{M}, \phi)$ forms a chart on $\mathbb{M}(m \times n, \mathbb{R})$, whose domain is whole of $\mathbb{M}$. Let us denote $\mathbb{A} = \{(\mathbb{M}, \phi)\}$ to be the collection of charts of $\mathbb{M}(m \times n, \mathbb{R})$. It only remains to show that all pairs of members in $\mathbb{A}$ are C^∞-compatible. For this, let $(\mathbb{M}, \phi) \in \mathbb{A}$ and $(\mathbb{M}, \psi) \in \mathbb{A}$. The transition functions

$$\phi \circ \psi^{-1} : \mathbb{R}^{mn} \to \mathbb{R}^{mn} \text{ and } \psi \circ \phi^{-1} : \mathbb{R}^{mn} \to \mathbb{R}^{mn}$$

are both identity maps and hence C^∞-compatible. So $\mathbb{A}$ forms a C^∞-atlas of $\mathbb{M}(m \times n, \mathbb{R})$. Thus, $\mathbb{M}(m \times n, \mathbb{R})$ form a smooth manifold of dimension mn.

In particular, if $m = n$ then $M(n \times n, \mathbb{R})$ **forms a smooth manifold of dimension** n^2.

Problem 2.17 Prove that the topological space $GL(n, \mathbb{R})$ forms a smooth manifold of dimension n^2.

Solution: 17 Here $GL(n, \mathbb{R}) = \{A \in M(n \times n, \mathbb{R}) : |A| \neq 0\}$. It is clear that the determinant map

$$\mathbb{D} : M(n \times n, \mathbb{R}) \to \mathbb{R}, \quad i.e. \ (a_{ij}) \mapsto |(a_{ij})|,$$

where $i, j = 1, 2, 3, \dots, n$ is continuous (also smooth). Hence, the inverse image of the open subset $R \setminus \{0\}$ of $\mathbb{R}$ is open in $M(n \times n, \mathbb{R})$, i.e. $\mathbb{D}^{-1}(R \setminus \{0\})(= GL(n, \mathbb{R}))$ is open in $M(n \times n, \mathbb{R})$. And $M(n \times n, \mathbb{R})$ being n^2-dimensional smooth manifold (refer to Problem 2.16), we can say that $GL(n, \mathbb{R})$ **is also** n^2**-dimensional smooth manifold** (refer to Example 2.6).

Problem 2.18 Give an example of a non-Hausdorff space having differentiable structures.

Solution: 18 Let $A \subset \mathbb{R}^2$ be such that

$$A = U \bigcup \{(0, 1)\},$$

where $U = \{(x, 0) | x \in \mathbb{R}\}$. Let $U_1 = \{(x, 1) | x \in \mathbb{R}\}$. We define $\phi : U \to \mathbb{R}$ by $\phi(x, 0) = x$ and $\psi : U_1 \to \mathbb{R}$ by

$$\psi(x, 1) = \begin{cases} 0, & x = 0 \\ x, & x \neq 0. \end{cases}$$

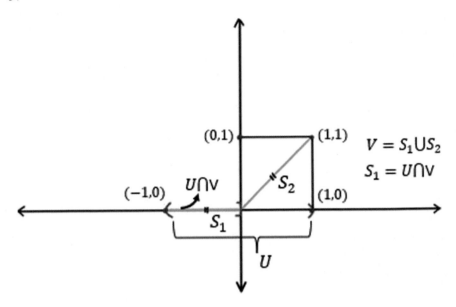

$V = S_1 \cup S_2$

$S_1 = U \cap V$

Fig. 2.8 Not atlas

Both ϕ and ψ are well-defined on U and U_1, respectively. Also ϕ and ψ are homeomorphisms and $U_1 \bigcup U_2 = A$. The transitions functions $\phi \circ \psi^{-1}$ and $\psi \circ \phi^{-1}$ are the identity functions on $\mathbb{R} \setminus \{0\}$. So $\{(U, \phi), (U_1, \psi)\}$ forms a C^∞ atlas on A.

Let V_1 be an open neighbourhood of $(0, 0)$ and V_2 be an open neighbourhood of $(0, 1)$ in A. Then both $\phi(U \bigcap V_1)$ and $\psi(U_1 \bigcap V_2)$ are open subsets of $\mathbb{R}$ containing 0. So $\exists\, a \neq 0$ such that $a \in \phi(U \bigcap V_1) \bigcap \psi(U_1 \bigcap V_2)$, which implies $(a, 0) \in V_1 \bigcap V_2$. Hence the topology of A is non-Hausdorff.

Problem 2.19 Consider $S = \{(x, 0) \in \mathbb{R}^2 : x \in (-1, 1)\} \cup \{(x, x) \in \mathbb{R}^2 : x \in (0, 1)\}$ where $U = \{(x, 0) : x \in (-1, 1)\}$, $\phi : U \to \mathbb{R}$ is such that $\phi(x, 0) = x$ and $V = \{(x, 0) : x \in (-1, 0]\} \cup \{(x, x) : x \in (0, 1)\}$, $\psi : V \to \mathbb{R}$ be such that $\psi(x, 0) = x$, $\psi(x, x) = x$. Is $\mathbb{A} = \{(U, \phi), (V, \psi)\}$ an atlas on S?

Solution: 19 Endow $S = U \cup V$ with the subspace topology inherited from $\mathbb{R}^2$. Let $S_1 = \{(x, 0) : x \in (-1, 0]\}$ and $S_2 = \{(x, x) : x \in (0, 1)\}$. Then $V = S_1 \cup S_2$. Here ϕ and ψ are both homeomorphisms respectively on U and V. Hence (U, ϕ) and (V, ψ) form the charts on S. Now the transition functions $\phi \circ \psi^{-1} : \psi(U \cap V) \to \phi(U \cap V)$ and $\psi \circ \phi^{-1} : \phi(U \cap V) \to \psi(U \cap V)$ are the identity map on $\phi(U \cap V)$ and $\psi(U \cap V)$ respectively, where $\phi(U \cap V) = (-1, 0] = \psi(U \cap V)$, which are not open in $\mathbb{R}$. Thus, $\mathcal{A}$ does not form an atlas on S (Fig. 2.8).

Problem 2.20 Let $U = \{(x, 0) : x \in \mathbb{R}\}$ and $V = \{(x, 0) : x < 0\} \cup \{(x, 1) : x > 0\}$. Suppose the maps $\phi : U \to \mathbb{R}$ and $\psi : V \to \mathbb{R}$ are defined respectively by $\phi(x, 0) = x$ and $\psi(x, 0) = x$, $\psi(x, 1) = x$. Prove that $\mathbb{A} = \{(U, \phi), (V, \psi)\}$ an atlas on $S = U \cup V$.

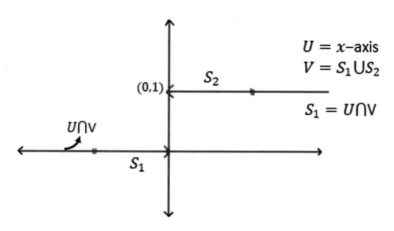

Fig. 2.9 Atlas

Solution: 20 Endow $S = U \cup V$ with the subspace topology inherited from $\mathbb{R}^2$. Let $S_1 = \{(x, 0) : x \in (-\infty, 0)\}$ and $S_2 = \{(x, x) : x \in (0, 1)\}$. Then $V = S_1 \cup S_2$. Here ϕ and ψ are both homeomorphisms respectively on U and V. Hence (U, ϕ) and (V, ψ) form the charts on S. Now the transition functions $\phi \circ \psi^{-1} : \psi(U \cap V) \to \phi(U \cap V)$ and $\psi \circ \phi^{-1} : \phi(U \cap V) \to \psi(U \cap V)$ are the identity map on $\phi(U \cap V)$ and $\psi(U \cap V)$ respectively, where $\phi(U \cap V) = (-\infty, 0) = \psi(U \cap V)$, which are open in $\mathbb{R}$. Also the map $\phi \circ \psi^{-1}$ and $\psi \circ \phi^{-1}$ are C^∞. Thus, $\mathcal{A}$ does form an atlas on S (Fig. 2.9).

Exercises

Exercise 2.7 *Give an example of a topological manifold which does not admit differential structure.*

Exercise 2.8 *Let $(M, \mathbb{A})$ be a smooth manifold. Prove that there exists a chart (U, ϕ) of $\mathbb{A}$ such that $\phi(p) = 0, \ p \in U$.*

Remark 2.18 The condition of second countability in the definition of smooth manifold implies paracompactness which further implies metric structure in the manifold. Since the present book is considering **only** the different aspects of smooth manifold, the condition is redundant here.

From now onwards, unless otherwise stated, **a manifold will mean a smooth manifold**.

2.4 Stereographic Projection

The stereographic projection is a particular mapping that projects a sphere onto a plane. The projection is defined on the entire sphere, except at the projection point.

To begin with, let us consider the sphere $S^1 = \{(x, y) : (x, y) \in \mathbb{R}^2, x^2 + y^2 = 1\}$. The stereographic projection, on the sphere S^1, from the North Pole $N = (0, 1)$(respectively, South Pole $S = (0, -1)$) onto the line $y = 0$ (i.e. x-axis) is the map which assigns any point $p \in S^1 \setminus \{N\}$(respectively, $p \in S^1 \setminus \{S\}$) to the point where the straight line through p and N(respectively, through S) intersects the line $y = 0$.

The inverse of the stereographic projection is the map from the x-axis to $S^1 \setminus \{N\}$ (respectively, $S^1 \setminus \{S\}$) assigning the point q in the line $y = 0$ to the point where the straight line through N (respectively, through S) and q intersect S^1.

Similarly, the stereographic projection, on the sphere $S^2 = \{(x, y, z) : (x, y, z) \in \mathbb{R}^3, x^2 + y^2 + z^2 = 1\}$, from the North Pole $N = (0, 0, 1)$(respectively, South Pole $S = (0, 0, -1)$) onto the plane $z = 0$ (i.e. xy-plane) is the map which assigns any point $p \in S^2 \setminus \{N\}$(respectively, $p \in S^2 \setminus \{S\}$) to the point where the straight line through p and N(respectively, through S) intersects the plane $z = 0$.

The inverse of the stereographic projection is the map from the xy-plane to $S^2 \setminus \{N\}$(respectively, $S^2 \setminus \{S\}$) assigning the point q in the plane $z = 0$ to the point where the straight line through N(respectively, through S) and q intersect S^2.

On generalization, we can define the stereographic projection and its inverse for the n-dimensional sphere $S^n = \{(x_1, x_2, \ldots, x_{n+1}) : (x_1, x_2, \ldots, x_{n+1}) \in \mathbb{R}^{n+1}, \sum_{i=1}^{n+1}(x_i)^2 = 1\}$ as follows.

The stereographic projection, on the sphere S^n, from the North Pole $N = (0, 0, \ldots, 0, 1)$ (respectively, South Pole $S = (0, 0, \ldots, 0, -1)$) onto the plane $x^{n+1} = 0$ is the map which assigns any point $p \in S^n - \{N\}$ (respectively, $p \in S^n - \{S\}$) to the point where the straight line through p and N(respectively, S) intersects the plane $x^{n+1} = 0$.

The inverse of the stereographic projection is the map from the plane $x^{n+1} = 0$ to $S^n \setminus \{N\}$ (respectively, $S^n \setminus \{S\}$) assigning the point q in the plane $x^{n+1} = 0$ to the point where the straight line through N(respectively, through S) and q intersect S^n.

Problem 2.21 Using stereographic projection with x-axis as the image line, show that S^1 is a smooth manifold (Fig. 2.10).

Solution: 21 Note that $N = (0, 1)$ and $S = (0, -1)$ are the North and South Poles of S^1. Let $p = (x, y) \in S^1$ and consider the set $U = S^1 \setminus \{N\}$ and $V = S^1 \setminus \{S\}$. From the definition of stereographic projection, $\phi : U \to \mathbb{R}$ and $\psi : V \to \mathbb{R}$ are given by

$$\phi(x, y) = \frac{x}{1 - y}, \quad \psi(x, y) = \frac{x}{1 + y}.$$

Fig. 2.10 Stereographic
projection on S^1

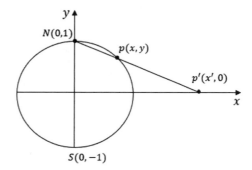

Our claim is that $\{(U, \phi), (V, \psi)\}$ forms an atlas of S^1. It is obvious that ϕ and ψ are homeomorphisms. Here $x^2 + y^2 = 1 \Rightarrow x^2 = (1 - y)(1 + y)$. Let $x' = \dfrac{x}{1 - y}$. Then by virtue of foregoing equation, we have

$$x'^2(1 - y)^2 = x^2 \Rightarrow (x')^2(1 - y) = (1 + y) \Rightarrow y = \frac{x'^2 - 1}{1 + x'^2}.$$

Similarly, $x = \dfrac{2x'}{1 + x'^2}$. Hence, the inverse map $\phi^{-1} : \mathbb{R} \to U$ is given by

$$\phi^{-1}(x') = (x, y) = \left(\frac{2x'}{1 + x'^2}, \frac{x'^2 - 1}{1 + x'^2} \right).$$

Here $\phi(U \cap V) = \mathbb{R} - \{0\}$ and $\psi(U \cap V) = \mathbb{R} - \{0\}$. Now, $\psi \circ \phi^{-1} : \phi(U \cap V) \to \psi(U \cap V)$ is given by $(\psi \circ \phi^{-1})(t) = \dfrac{1}{t}$, is a C^∞ function. Similarly, $\phi \circ \psi^{-1}$ is also C^∞ function. This proves $\{(U, \phi), (V, \psi)\}$ forms a C^∞ atlas of S^1. Hence S^1 is a smooth manifold.

Problem 2.22 Using stereographic projection with an equatorial plane as the image plane, prove that S^2 is a smooth manifold (Fig. 2.11).

Solution: 22 Note that $N = (0, 0, 1)$ and $S = (0, 0, -1)$ are the North and South Poles of S^2. Let $p = (x, y, z) \in S^2$ and consider the set $U = S^2 - \{N\}$ and $V = S^2 \setminus \{S\}$. Here $\phi : U \to \mathbb{R}^2$ and $\psi : V \to \mathbb{R}^2$ are given by

$$\phi(x, y, z) = \left(\frac{x}{1 - z}, \frac{y}{1 - z} \right), \quad \psi(x, y, z) = \left(\frac{x}{1 + z}, \frac{y}{1 + z} \right).$$

Our claim is that $\{(U, \phi), (V, \psi)\}$ forms an atlas of S^2.

It is clear that ϕ and ψ are homeomorphisms. Moreover, the inverse map $\phi^{-1} :$ $\mathbb{R}^2 \to U$ is given by

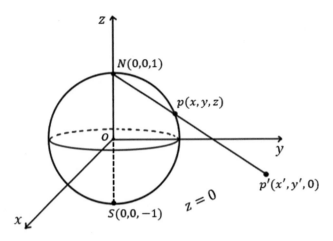

Fig. 2.11 Stereographic projection on S^2

$$\phi^{-1}(x', y') = (x, y, z) = \left(\frac{2x'}{1 + x'^2 + y'^2}, \frac{2y'}{1 + x'^2 + y'^2}, \frac{x'^2 + y'^2 - 1}{1 + x'^2 + y'^2} \right).$$

Note that $\phi(U \cap V) = \mathbb{R}^2 \setminus \{(0, 0)\}$ and $\psi(U \cap V) = \mathbb{R}^2 - \{(0, 0)\}$. Now, $\psi \circ \phi^{-1}$: $\phi(U \cap V) \to \psi(U \cap V)$ given by

$$(\psi \circ \phi^{-1})(x', y') = \left(\frac{x'}{x'^2 + y'^2}, \frac{y'}{x'^2 + y'^2} \right)$$

is a C^∞ function. Similarly, $\phi \circ \psi^{-1}$ is also C^∞ function. This proves $\{(U, \phi),$ $(V, \psi)\}$ forms a C^∞ atlas of S^2. Hence S^2 is a smooth manifold.

Problem 2.23 Show that the n-dimensional sphere S^n is a smooth manifold.

Solution: 23 Here $N = (\underbrace{0, 0, 0, \ldots, 0}_{n-\text{zeroes}}, 1)$ and $S = (\underbrace{0, 0, 0, \ldots, 0}_{n-\text{zeroes}}, -1)$ are the North and South Poles of S^n. Note that

$$U = S^n \setminus \{(\underbrace{0, 0, 0, \ldots, 0}_{n-\text{zeroes}}, 1)\}, \quad V = S^n - \{(\underbrace{0, 0, 0, \ldots, 0}_{n-\text{zeroes}}, -1)\}$$

and $\phi : U \to \mathbb{R}^n$ and $\psi : V \to \mathbb{R}^n$ are given by

$$\phi(x_1, x_2, \ldots, x_{n+1}) = \left(\frac{x_1}{1 - x_{n+1}}, \frac{x_2}{1 - x_{n+1}}, \ldots, \frac{x_n}{1 - x_{n+1}} \right),$$

$$\psi(x_1, x_2, \ldots, x_{n+1}) = \left(\frac{x_1}{1 + x_{n+1}}, \frac{x_2}{1 + x_{n+1}}, \ldots, \frac{x_n}{1 + x_{n+1}} \right).$$

The inverse map $\phi^{-1} : \mathbb{R}^n \to U$ is

$$\phi^{-1}(x'_1, x'_2, \ldots, x'_n) = \left(\frac{2x'_1}{1 + \sum_i x_i'^2}, \frac{2x'_2}{1 + \sum_i x_i'^2}, \cdots, \frac{2x'_n}{1 + \sum_i x_i'^2}, \frac{\sum_i x_i'^2 - 1}{1 + \sum_i x_i'^2} \right).$$

So, $\psi \circ \phi^{-1} : \phi(U \cap V) \to \psi(U \cap V)$ is given by

$$(\psi \circ \phi^{-1})(x'_1, x'_2, \ldots, x'_n) = \left(\frac{x'_1}{\sum_i (x_i)^2}, \frac{x'_2}{\sum_i (x_i)^2}, \cdots, \frac{x'_n}{\sum_i (x_i)^2} \right),$$

which is C^∞ in $\phi(U \cap V) = \mathbb{R}^n - \{(0, 0, 0 \ldots, 0, 0)\}$. Hence $\{(U, \phi), (V, \psi)\}$ forms a C^∞ atlas of S^n. So S^n is a smooth manifold.

2.5 Orientable Surface

A regular surface S is said to be orientable if there exists a family of surface patches (coordinate neighbourhoods), which will cover S, in such a way that if a point p of S belongs to the intersection of two surface patches of the family, then the transition map between the surface patches of the family has positive Jacobian. The existence of such a family of surface patches is called an orientation of S and S is called an orientable surface. If such a family does not exist, the surface is called non-orientable.

$S^n (n > 1)$ forms an example of orientable surface. The sphere can be covered by two coordinate neighbourhoods (using stereographic projection; refer to examples on stereographic projection). If W is the intersection of these coordinate neighbourhoods, W is connected. Let $p \in W$ be fixed. Since the Jacobian is positive at $p \in W$, it follows from the connectedness of W that the Jacobian is everywhere positive. Hence $S^n (n > 1)$ is orientable.

Now, we are going to focus our attention on an example of a non-orientable surface, the so-called Möbius Strip. From the geometrical point of view, a Möbius Band has the property that a figure moving around on the surface can come back to its starting point and transform into its mirror image, so it is impossible to decide consistently which of the two possible rotational directions on the surface to call **clockwise** and which **counterclockwise**, or which is the **front** and which is the **back** side. Let us denote the Möbius band by M (Fig. 2.12).

Fig. 2.12 Möbius band

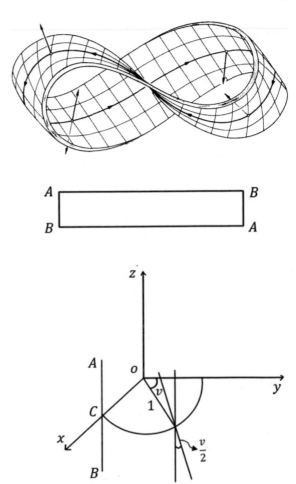

The Möbius band is obtained by rotating an open segment AB around its midpoint $C = (1, 0, 0)$ at the same time as C moves around a circle $S^1 \equiv x^2 + y^2 = 1$ in such a manner that as C moves once around S^1, the segment AB makes a half turn around C. After C has rotated by an angle v around the z-axis, AB should have rotated by $\dfrac{v}{2}$ around C in the plane containing C and the z-axis. Initially, the point of the segment AB is at $(1, 0, u)$.

Let $U_1 = \{(u, v) : (u, v) \in \mathbb{R}^2, -\frac{1}{2} < u < \frac{1}{2}, 0 < v < 2\pi\}$ and $U_2 = \{(\tilde{u}, \tilde{v}) : (\tilde{u}, \tilde{v}) \in \mathbb{R}^2, -\frac{1}{2} < \tilde{u} < \frac{1}{2}, -\pi < \tilde{v} < \pi\}$. Each of U_1 and U_2 are open in $\mathbb{R}^2$. We define $\phi : U_1 \to M$ and $\psi : U_2 \to M$ by

$$\phi(u, v) = \left(\left(1 - u \sin \frac{v}{2}\right) \cos v, \left(1 - u \sin \frac{v}{2}\right) \sin v, u \cos \frac{v}{2}\right)$$

$$\psi(\tilde{u}, \tilde{v}) = \left(\left(1 - \tilde{u} \sin \frac{\tilde{v}}{2}\right) \cos \tilde{v}, \left(1 - \tilde{u} \sin \frac{\tilde{v}}{2}\right) \sin \tilde{v}, \tilde{u} \cos \frac{\tilde{v}}{2}\right).$$

Here, $\phi(U_1 \cap U_2)$ is not connected but consists of two connected components given by

$$W_1 = \{\phi(u, v) : (u, v) \in \mathbb{R}^2, -\frac{1}{2} < u < \frac{1}{2}, 0 < v < \pi\},$$

$$W_2 = \{\phi(u, v) : (u, v) \in \mathbb{R}^2, -\frac{1}{2} < u < \frac{1}{2}, \pi < v < 2\pi\}.$$

Hence, $\phi(U_1 \cap U_2) = W_1 \cup W_2$. Geometrically, $\phi(U_1 \cap U_2)$ is the union of the rectangles given by $0 < v < \pi$ and $\pi < v < 2\pi$, with $-\frac{1}{2} < u < \frac{1}{2}$. If $0 < v < \pi$, then it is clear that $(\phi^{-1} \circ \psi)(u, v) = (u, v)$. If $\pi < v < 2\pi$, we have $v - \tilde{v} = 2\pi$. Now combining $\sin \frac{\tilde{v}}{2} = -\sin \frac{v}{2}$, $\cos \frac{\tilde{v}}{2} = -\cos \frac{v}{2}$ and $\phi(u, v) = \psi(\tilde{u}, \tilde{v})$ implies $\tilde{u} = -u$. Hence

$$(\phi^{-1} \circ \psi)(u, v) = (u, v), \text{ if } 0 < v < \pi$$
$$= (-u, v - 2\pi), \text{ if } \pi < v < 2\pi.$$

Hence, $\phi^{-1} \circ \psi$ forms the transition map between the two surface patches $\phi(u, v)$ and $\psi(u, v)$ for M. This proves $\{(U_1, \phi), (U_2, \psi)\}$ forms an C^∞ atlas for M, hence a smooth manifold of dimension 2. Here

$$\phi_u\big|_{u=0} = \left(-\sin \frac{v}{2} \cos v, -\sin \frac{v}{2} \sin v, \cos \frac{v}{2}\right),$$

$$\phi_v\big|_{u=0} = (-\sin v, \cos v, 0).$$

Therefore, $\phi_u\big|_{u=0} \times \phi_v\big|_{u=0} = (-\cos v \cos \frac{v}{2}, -\sin v \cos \frac{v}{2}, -\sin \frac{v}{2})$. Now the unit normal is given by $N_\phi = \frac{\phi_u \times \phi_v}{\|\phi_u \times \phi_v\|} = \phi_u \times \phi_v$. If possible, let us assume M to be orientable. Then $\exists$ a well-defined unit normal vector $N : M \to \mathbb{R}^3$ at every point of M, which varies smoothly over M. At $\phi(0, v) \in S^1$, we have $N = \mu(v)N_\phi$ where $\mu : (0, 2\pi) \to \mathbb{R}$ is smooth. Also, $\mu(v) = \pm 1 \; \forall \; v$. It follows that

$$\mu(v) = +1, \; \forall \; v \in (0, 2\pi) \text{ or } \mu(v) = -1, \; \forall \; v \in (0, 2\pi).$$

Depending on the changes $u \to v, \tilde{u} \to \tilde{v}$ has to be made, suppose $\mu = 1$. As N is smooth, at $\phi(0, 0) = \phi(0, 2\pi)$, we have

$$N = \lim_{v \to 0} N_\phi = (-1, 0, 0)$$

$$N = \lim_{v \to 2\pi} N_\phi = (1, 0, 0).$$

This leads to a contradiction, which proves M is non-orientable.

Remark 2.19 The Jacobian(J) of $\phi^{-1} \circ \psi$, i.e. $J(\phi^{-1} \circ \psi) = \dfrac{\partial(u, v)}{\partial(\tilde{u}, \tilde{v})} = 1 > 0$ in

W_1 and $J(\phi^{-1} \circ \psi) = \dfrac{\partial(u, v)}{\partial(\tilde{u}, \tilde{v})} = -1 < 0$ in W_2.

2.6 Product Manifold

Let M_1^m and M_2^n be smooth manifolds, with differentiable structures $\mathbb{A}_1$ and $\mathbb{A}_2$, of dimensions m and n, respectively. First we prove that $M_1 \times M_2$, with the product topology, is a $(m + n)$-dimensional topological manifold.

Since M_1 and M_2 is a smooth manifold, the topology of M_1 and M_2 is Hausdorff and second countable. Therefore, the product topology of $M_1 \times M_2$ is Hausdorff and second countable.

Let $(x_1, x_2) \in M_1 \times M_2$. Since $x_1 \in M_1$ and M_1 is a m-dimensional smooth manifold with differentiable structure $\mathbb{A}_1$, $\exists$ a coordinate chart $(U, \phi) \in \mathbb{A}_1$ such that $x_1 \in U$. Similarly, $\exists$ a coordinate chart $(V, \psi) \in \mathbb{A}_2$ such that $x_2 \in V$. Hence, ϕ is a homeomorphism from the open subset U of M_1 onto the open subset $\phi(U)$ of R^m. Similarly, ψ is a homeomorphism from the open subset V of M_2 onto the open subset $\psi(V)$ of R^n. Since U is an open neighbourhood of x_1 and V is an open neighbourhood of x_2, the Cartesian product $U \times V$ is an open neighbourhood of $(x_1, x_2) \in M_1 \times M_2$. Furthermore, the Cartesian product $\phi(U) \times \psi(V)$ is open in $\mathbb{R}^m \times \mathbb{R}^n \equiv \mathbb{R}^{m+n}$. Let us define a function

$$\phi \times \psi : U \times V \to \phi(U) \times \psi(V), \ (x_1, x_2) \mapsto (\phi(x_1), \psi(x_2)), \ \forall \, (x_1, x_2) \in U \times V.$$

Then $\phi \times \psi$ is homeomorphic to an open subset $\phi(U) \times \psi(V)$ in $\mathbb{R}^{m+n}$. Hence, $M_1 \times M_2$, with the product topology, is a $(m + n)$-dimensional topological manifold.

Let us denote the collection of coordinate charts of $M_1 \times M_2$ by $\mathcal{C}$. Then

$$\mathcal{C} = \{(U \times V, \phi \times \psi) : (U, \phi) \in \mathbb{A}_1, (V, \psi) \in \mathbb{A}_2\}.$$

Here $\bigcup \{U \times V : (U \times V, \phi \times \psi) \in \mathcal{C}\} = M_1 \times M_2$. To show that $M_1 \times M_2$ is a $(m + n)$-dimensional smooth manifold, it only remains to show that all pairs of members in $\mathcal{C}$ are C^k-compatible, for every $k = 1, 2, 3, \ldots, \ldots$.

Let $(U \times V, \phi \times \psi) \in \mathcal{C}$ and $(\tilde{U} \times \tilde{V}, \tilde{\phi} \times \tilde{\psi}) \in \mathcal{C}$. Then the transition function

$$(\phi \times \psi) \circ (\tilde{\phi} \times \tilde{\psi})^{-1} = (\phi \circ \tilde{\phi}^{-1}) \times (\psi \circ \tilde{\psi}^{-1}) : (\tilde{\phi} \times \tilde{\psi})((U \times V) \cap (\tilde{U} \times \tilde{V})) \to (\phi \times \psi)((U \times V) \cap (\tilde{U} \times \tilde{V}))$$

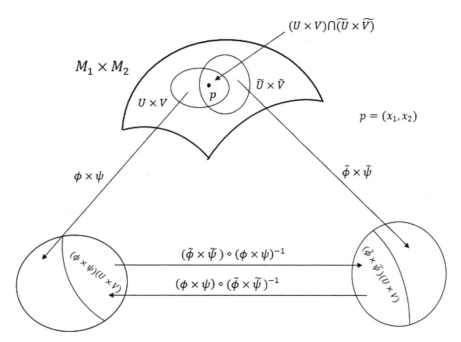

Fig. 2.13 Product manifold

is C^k, as both $(\phi \circ \tilde{\phi}^{-1})$ and $(\psi \circ \tilde{\psi}^{-1})$ are C^k compatible. By similar reasoning, the transition function

$$(\psi \times \phi) \circ (\tilde{\psi} \times \tilde{\phi})^{-1} = (\psi \circ \tilde{\psi}^{-1}) \times (\phi \circ \tilde{\phi}^{-1}) : (\tilde{\psi} \times \tilde{\phi})\big((U \times V) \cap (\tilde{U} \times \tilde{V})\big) \to (\psi \times \phi)\big((U \times V) \cap (\tilde{U} \times \tilde{V})\big)$$

is also so. Hence, $\mathcal{C}$ forms a C^∞ atlas on $M_1 \times M_2$, so it is a smooth manifold of dimension $(m + n)$ (Fig. 2.13).

Example 2.7 Since S^1 is a smooth manifold, the torus $S^1 \times S^1$ is a smooth manifold.

Problem 2.24 Prove that the infinite cylinder is a smooth manifold.

Solution: 24 Since the infinite cylinder can be expressed as a product of smooth manifolds S^1 and $\mathbb{R}$, it is a smooth manifold.

Exercise

Exercise 2.9 Prove that the n-dimensional torus is a smooth manifold.

2.7 Smooth Function on Smooth Manifold

Let M be a smooth manifold of dimension n (Fig. 2.14).

Let us recall the definition of a smooth function on a smooth manifold. A function $f : M \to \mathbb{R}$ is said to be of class C^∞ or **smooth at a point** $p \in M$, if for every chart (U, ϕ) of p in M, the function

$$f \circ \phi^{-1} : \phi(U) \subset \mathbb{R}^n \to \mathbb{R}$$

is of class C^∞ at $\phi(p)$. The function f is said to be C^∞ on M if it is C^∞ at every point of M.

Proposition 2.3 *The notion of smoothness of a map is independent of the choice of a coordinate chart.*

Proof Let M be an n-dimensional smooth manifold and let $f : M \to \mathbb{R}$ be smooth at every point of M. Hence, for every chart (U, ϕ) of $p \in M$, $f \circ \phi^{-1} : \phi(U) \subset \mathbb{R}^n \to \mathbb{R}$ is of class C^∞ at $\phi(p)$. If (V, ψ) is any other chart of p, where $p \in U \cap V$, then on $\psi(U \cap V)$
$$f \circ \psi^{-1} = (f \circ \phi^{-1}) \circ (\phi \circ \psi^{-1}).$$

Let $\mathcal{A} = \{(U_\alpha, \phi_\alpha)\}$ be an atlas of M, where each $(U, \phi), (V, \psi) \in \mathcal{A}$. Thus, each $\phi \circ \psi^{-1}, \psi \circ \phi^{-1}$ is of class C^∞ and hence from above, $f \circ \psi^{-1}$ is C^∞ on $\psi(U \cap V)$.

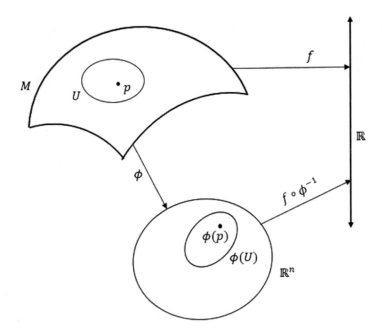

Fig. 2.14 Smooth function

Consequently, the smoothness of f on M is independent of the choice of any coordinate chart on M. This completes the proof.

Proposition 2.4 *Let M be an n-dimensional smooth manifold and $f : M \to \mathbb{R}$ be a map. Then the followings are equivalent:*

(i) *The map $f : M \to \mathbb{R}$ is of class C^∞.*
(ii) *$f \circ \phi^{-1} : \phi(U) \subset \mathbb{R}^n \to \mathbb{R}$ is of class C^∞, for every chart $(U, \phi) \in \mathcal{A}$, $\mathcal{A}$ being the atlas.*
(iii) *$f \circ \psi^{-1} : \psi(V) \subset \mathbb{R}^n \to \mathbb{R}$ is of class C^∞, for every chart $(V, \psi) \in \mathcal{A}$, $\mathcal{A}$ being the atlas.*

Proof **(ii)⇒(i):** Let (ii) hold. Then $f = (f \circ \phi^{-1}) \circ \phi : M \to \mathbb{R}$ is of class C^∞ at $\mathbb{R}$, as (U, ϕ) is given to be a chart. Thus (ii)⇒(i).
(i)⇒(iii): Let (i) hold. Then by definition, $\exists$ a chart (U, ϕ) at $p \in M$ such that $f \circ \phi^{-1} : \phi(U) \subset \mathbb{R}^n \to \mathbb{R}$ is of class C^∞ at $\phi(p)$.
If (V, ψ) is any chart at p on M, $p \in U \cap V$, then

$$f \circ \psi^{-1} = (f \circ \phi^{-1}) \circ (\phi \circ \psi^{-1})$$

is of class C^∞ on $\psi(V)$. Thus (i)⇒(iii).
(iii)⇒(ii): Let (iii) hold. Let $\mathcal{A}$ be an atlas on M, where $(U, \phi), (V, \psi) \in \mathcal{A}$. By definition, $\phi \circ \psi^{-1}$ and $\psi \circ \phi^{-1}$ are C^∞-related. Thus

$$f \circ \phi^{-1} = (f \circ \psi^{-1}) \circ (\psi \circ \phi^{-1})$$

is of class C^∞ on $\phi(U)$.

We shall often denote by $F(M)$, the set of all C^∞-functions on M and will sometime denote by $F(p)$, all the C^∞-functions at p of M. It is to be noted that such $F(M)$ is

(i) an algebra over $\mathbb{R}$
(ii) a module over $\mathbb{R}$,

where the defining relations are

$$\begin{cases} (a)\ (f + g)(p) = f(p) + g(p), & ; \\ (b)\ (fg)(p) = f(p)g(p), & ; \\ (c)\ (\lambda f)(p) = \lambda f(p),\ \forall\ f, g \in F(M),\ \lambda \in \mathbb{R}. \end{cases}$$

Now we are going to discuss smooth functions between smooth manifolds.
Let M be an n-dimensional and N be an m-dimensional manifold (Fig. 2.15).
A mapping $f : M \to N$ is said to be a **differentiable mapping** of class C^k if for every chart (U, ϕ) of $p \in M^n$ and every chart (V, ψ) of $f(p) \in N^m$,

$$\begin{cases} (i)\ f(U) \subset V \\ (ii)\ \psi \circ f \circ \phi^{-1} : \phi(U) \subset \mathbb{R}^n \to \psi(V) \subset \mathbb{R}^m,\ \text{is of class } C^k. \end{cases} \tag{2.9}$$

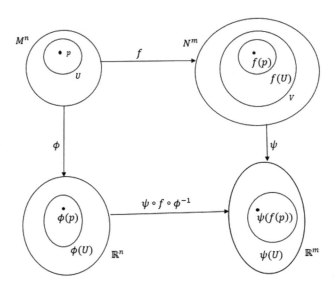

Fig. 2.15 Differentiable mapping

By differentiable mapping, we shall mean unless otherwise stated, a mapping of class C^∞.

Let M and N be two n-dimensional smooth manifolds. A mapping $f : M \to N$ is said to be a **diffeomorphism** if

$$\begin{cases} (i) \ f \text{ is a differentiable mapping.} \\ (ii) \ f \text{ is a bijection} \\ (iii) \ f^{-1} \text{ is of class } C^\infty. \end{cases} \qquad (2.10)$$

In such cases, M, N are said to be diffeomorphic to each other. A diffeomorphism f of M onto itself is called a **transformation** on M.

Proposition 2.5 *The notion of smoothness of a map between two smooth manifolds is independent of the choice of a coordinate chart.*

Proof Let M be an n-dimensional and N be an m-dimensional manifold. Let us consider the map $f : M \to N$ to be differentiable of class C^∞. Note that (U, ϕ) and (V, ψ) are any charts about $p \in M$ and $f(p) \in N$ respectively with $f(U) \subset V$ such that the map $\psi \circ f \circ \phi^{-1}$ being C^∞. Then for $p \in M$, suppose there exists charts $(\tilde{U}, \tilde{\phi})$ of p and $(\tilde{V}, \tilde{\psi})$ of $f(p)$ with $f(\tilde{U}) \subset \tilde{V}$ such that the map

$$\tilde{\psi} \circ f \circ \tilde{\phi}^{-1} : \tilde{\phi}(\tilde{U}) \subset \mathbb{R}^n \to \tilde{\psi}(\tilde{V}) \subset \mathbb{R}^m, \text{ is of class } C^\infty \text{ at } \tilde{\phi}(p).$$

However, $\tilde{\phi} \circ \phi^{-1}$ and $\psi \circ \tilde{\psi}^{-1}$ are C^∞ on open subset of Euclidean spaces. Hence,

$$(\psi \circ \tilde{\psi}^{-1}) \circ (\tilde{\psi} \circ f \circ \tilde{\phi}^{-1}) \circ (\tilde{\phi} \circ \phi^{-1}) = \psi \circ f \circ \phi^{-1}$$

is C^∞. This proves that the C^∞ structure of a map does not depend on any coordinate chart chosen.

Proposition 2.6 *If (U, ϕ) is a chart on a manifold M of dimension n, then the map $\phi : U \subset M \to \phi(U) \subset \mathbb{R}^n$ is a diffeomorphism.*

Proof Note that (U, ϕ) being a chart, ϕ is a homeomorphism.

Further, we use the atlas $\{(U, \phi)\}$ with a single chart on U and the atlas $\{\phi(U), I_{\phi(U)}\}$ with a single chart on $\phi(U)$. Then $I_{\phi(U)} \circ \phi \circ \phi^{-1} : \phi(U) \subset \mathbb{R}^n \to \phi(U)$ is the identity map. Thus, $I_{\phi(U)} \circ \phi \circ \phi^{-1}$ is of class C^∞ and by (2.9), ϕ is a differentiable mapping. Furthermore, $\phi \circ \phi^{-1} \circ I_{\phi(U)} : \phi(U) \to \phi(U)$ is an identity map, hence is of class C^∞. Thus, all the conditions of (2.10) are satisfied by ϕ and so is diffeomorphism.

Proposition 2.7 *Let $U \subset M$ be an open subset of a manifold M of dimension n. If $\phi : U \subset M \to \phi(U) \subset \mathbb{R}^n$ is a diffeomorphism, onto an open subset of $\mathbb{R}^n$, then (U, ϕ) is a chart on M.*

Proof For any chart (U, ϕ_α) on M, both ϕ_α and ϕ_α^{-1} are of class C^∞ (refer to Proposition 2.6).

Now ϕ being a diffeomorphism, it is C^∞. Consequently, the composite mappings $\phi \circ \phi_\alpha^{-1}$ and $\phi_\alpha \circ \phi^{-1}$ are of class C^∞. Hence (U, ϕ) is compatible with an atlas on M. Thus (U, ϕ) is a chart on M.

Problem 2.25 If $(x^1, x^2, \dots, x^n)$ and $(y^1, y^2, \dots, y^m)$ are respectively the local coordinate systems defined in the neighbourhood U of $p \in M^n$ and V of $f(p) \in N^m$, then it can be shown that

$$y^j \circ f = g^j(x^1, x^2, \dots, x^n), \quad \text{where} \tag{2.11}$$

$$g^j(q) = (\psi \circ f \circ \phi^{-1})(q), \quad q \in \phi(U). \tag{2.12}$$

Solution: 25 Let $\phi(p) = q$, $p \in U \subset M^n$. Then

$$g(\phi(p)) = (\psi \circ f \circ \phi^{-1})(\phi(p)) = (\psi \circ f)(p)$$
$$\text{or } u^j(g(\phi(p))) = u^j(\psi(f(p))) = (u^j \circ \psi)(f(p))$$
$$\text{or } g^j(\phi(p)) = y^j(f(p)) \text{ by } (2.1), (2.3)$$
$$\text{or } g^j(x^1(p), \dots, x^n(p)) = (y^j \circ f)(p) \text{ by } (2.2)$$
$$\text{i.e. } y^j \circ f = g^j(x^1, x^2, \dots, x^n).$$

Fig. 2.16 Differentiable
map between punctured
sphere and cylinder

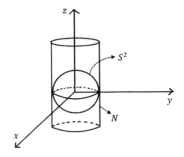

Problem 2.26 Obtain a differentiable map between the punctured sphere at two points $(0, 0, 1)$, $(0, 0, -1)$ and the cylinder N with infinite ends (Fig. 2.16).

Solution: 26 Let us consider the punctured sphere $M = S^2 - \{(0, 0, 1), (0, 0, -1)\}$ where $S^2 = \{(x, y, z) : (x, y, z) \in \mathbb{R}^3, x^2 + y^2 + z^2 = 1\}$. Here, the coordinate neighbourhood (U, ϕ) is given by $U = S^2 - \{(0, 0, 1), (0, 0, -1)\}$, $\phi : U \to \mathbb{R}^2$, where $\phi(\cos v \cos u, \cos v \sin u, \sin v) = (x, y)$. Therefore, $\phi^{-1} : \mathbb{R}^2 \to U$ is defined by $\phi^{-1}(x, y) = (\cos v \cos u, \cos v \sin u, \sin v)$.

Let us consider the cylinder $N = \{(\tilde{x}, \tilde{y}, \tilde{z}) : (\tilde{x}, \tilde{y}, \tilde{z}) \in \mathbb{R}^3, \tilde{x}^2 + \tilde{y}^2 = 1, 0 < \tilde{z} < 1\}$. Here, the coordinate neighbourhood (V, ψ) is given by $V = N$ and $\psi : N \to \mathbb{R}^2$, where $\psi(\cos u, \sin u, z) = (x, z)$. Therefore, $\psi^{-1} : \mathbb{R}^2 \to V$ is defined by $\psi^{-1}(x, z) = (\cos u, \sin u, z)$.

Let us define a map $f : M \to N$ by $(x, y, z) \mapsto (\tilde{x}, \tilde{y}, z)$. In other words, we can say that the line joining p and $f(p)$ is parallel to the xy-plane and orthogonal to the z-axis. Since the point $(\cos v \cos u, \sin v \sin u, \sin v)$ is moving from the sphere M parallel to the xy-plane and is orthogonal to the z-axis, therefore the x and y components of $(\cos v \cos u, \sin v \sin u, \sin v)$ will take the coordinate of the cylinder N but the z-component, i.e. $\sin v$ will remain unchanged. Hence,

$$f(\phi^{-1}(x, y)) = (\cos u, \sin u, \sin v) = \psi^{-1}(x, \sin v)$$
$$i.e. \ (\psi \circ f \circ \phi^{-1})(x, y) = (x, \sin v),$$

which shows $\psi \circ f \circ \phi^{-1}$ is differentiable.

Exercise

Exercise 2.10 *Let M and N be two smooth manifolds with $M = N = \mathbb{R}$. Let (U, ϕ) and (V, ψ) be two charts on M and N respectively, where $U = \mathbb{R}$, $\phi : U \to \mathbb{R}$ is the identity mapping and $V = \mathbb{R}$, $\psi : V \to \mathbb{R}$ is the mapping defined by $\psi(x) = x^3$. Show that the two structures defined on $\mathbb{R}$ are not C^∞-related even though M and N are diffeomorphic where $f : M \to N$ is defined by $f(t) = t^{1/3}$.*

2.8 Differential Curve and Tangent Vector

We are now in a position to introduce one of the important concepts of geometry, i.e. tangent vector. Geometers prefer to define the tangent vector at a point with respect to a curve. Hence, at first we shall define a curve on a manifold.

A differentiable curve at p on a manifold M is a differentiable mapping $\sigma : [a, b] \subset \mathbb{R} \to M^n$ such that $\sigma(t_0) = p$, $a \le t_0 \le b$ (Fig. 2.17).

Then by (1.1) and (2.1), we obtain

$$(x^i \circ \sigma)(t) = u^i(\phi(\sigma(t))) = u^i(\sigma^1(t), \sigma^2(t), \dots, \sigma^n(t)) = \sigma^i(t). \qquad (2.13)$$

Often, we write it as

$$x^i(t) = \sigma^i(t). \qquad (2.14)$$

The **tangent vector** to the curve $\sigma(t)$ at p of M is a function $X_p : F(p) \to \mathbb{R}$ defined by

$$\begin{cases} X_p f = \left[\lim_{h \to 0} \dfrac{f(\sigma(t+h)) - f(\sigma(t))}{h} \right]_{t=t_0} = \dfrac{d}{dt} f(\sigma(t)) \Big|_{t=t_0}, \ \forall \ f \in F(p) \\ \sigma(t_0) = p. \end{cases}$$

$$(2.15)$$

Note that

$$X_p(af + bg) = \frac{d}{dt}(af + bg)(\sigma(t)) \Big|_{t=t_0}, \ \forall \ f, g \in F(p), a, b \in \mathbb{R}; \ \text{by (2.15)}$$

$$= a\frac{d}{dt} f(\sigma(t)) \Big|_{t=t_0} + b\frac{d}{dt} g(\sigma(t)) \Big|_{t=t_0}, \ \text{by (1.8)}$$

$$X_p(af + bg) = a(X_p f) + b(X_p g), \ \text{by (2.15)}. \qquad (2.16)$$

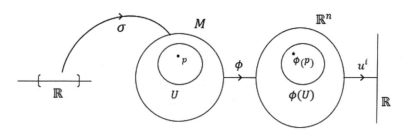

Fig. 2.17 Differential curve

Again

$$X_p(fg) = \frac{d}{dt}(fg)(\sigma(t))\big|_{t=t_0}, \text{ by (2.15)}$$

$$= \frac{d}{dt}f(\sigma(t))g(\sigma(t))\big|_{t=t_0}, \text{ by (1.8)}$$

$$= \frac{d}{dt}f(\sigma(t))\big|_{t=t_0}g(\sigma(t_0)) + f(\sigma(t_0))\frac{d}{dt}g(\sigma(t))\big|_{t=t_0}, \text{ by (1.9)}$$

$$X_p(fg) = (X_p f)g(p) + f(p)X_p g. \tag{2.17}$$

Equations (2.16) and (2.17) are respectively known as **linearity property** and **Leibnitz Product Rule**. Thus, the tangent vector at a point on a manifold is a derivation at that point.

Let $T_p(M)$ denote the set of all tangent vectors at p of M. We define

$$\begin{cases} (X_p + Y_p)f = X_p f + Y_p f, \ \forall \ X_p, Y_p \in T_p(M) \\ (\lambda X_p)f = \lambda(X_p f), \ \lambda \in \mathbb{R}. \end{cases} \tag{2.18}$$

Clearly, $T_p(M)$ is a real vector space (refer to any standard textbook of Linear Algebra). Hence, $T_p(M)$ must have a basis.

If $(x^1, x^2, \ldots, x^n)$ is the local coordinate system in a neighbourhood U of $p \in M$, then for each $i = 1, 2, \ldots, n$, we define a mapping $\frac{\partial}{\partial x^i} : F(p) \to \mathbb{R}$ by

$$\left(\frac{\partial}{\partial x^i}\right)_p f = \left(\frac{\partial f}{\partial x^i(t)}\right)(p), \ \forall \ f \in F(p). \tag{2.19}$$

Clearly, each $\left(\frac{\partial}{\partial x^i}\right)_p : i = 1, 2, 3, \ldots, n$ satisfies (2.16) and (2.17).

Let us define a differentiable curve $\sigma : [a, b] \subset \mathbb{R} \to M$ by

$$\begin{cases} \sigma^i(t) = \sigma^i(t_0) \text{ for fixed } i \\ \sigma^j(t) = 0, \ j = 1, 2, 3, \ldots, i-1, i+1, \ldots, n. \end{cases} \tag{2.20}$$

Then

$$\frac{d}{dt}f(\sigma(t))\big|_{t=t_0} = \sum_i \frac{\partial f(\sigma(t))}{\partial \sigma^i(t)} \frac{d\sigma^i(t)}{dt}\big|_{t=t_0}, \text{ by chain rule}$$

$$= \frac{\partial f(\sigma(t_0))}{\partial \sigma^i(t)}\big|_{t=t_0}, \text{ for fixed } i, \text{ by (2.20)}$$

$$= \frac{\partial f(p)}{\partial x^i(t)}, \text{ by (2.14)}$$

$$= \left(\frac{\partial}{\partial x^i} \right)_p f, \text{ by (2.19)}.$$

Thus, each $\left(\frac{\partial}{\partial x^i} \right)_p : i = 1, 2, 3, \ldots, n$ is a tangent vector to the curve σ defined by (2.20). Further from (2.15), we have

$$X_p f = \frac{d}{dt} f(\sigma(t)) \bigg|_{t=t_0}$$

$$= \sum_i \frac{\partial f(\sigma(t_0))}{\partial x^i(t)} \left(\frac{dx^i(t)}{dt} \right) \bigg|_{t=t_0}$$

$$= \sum_i \left(\frac{dx^i(t)}{dt} \right) \bigg|_{t=t_0} \left(\frac{\partial}{\partial x^i} \right)_p f$$

$$= \sum_i \xi^i(p) \left(\frac{\partial}{\partial x^i} \right)_p f, \text{ say, where}$$

$$\begin{cases} \xi^i(p) = \left(\frac{dx^i(t)}{dt} \right), & i = 1, 2, 3, \ldots, n \\ \xi^i : M \to \mathbb{R}, \text{ are differentiable functions on } M. \end{cases} \tag{2.21}$$

Thus we write

$$X_p = \sum \xi^i(p) \left(\frac{\partial}{\partial x^i} \right)_p, \quad \forall f \in F(p). \tag{2.22}$$

Finally, if we assume that $\sum \xi^i(p) \left(\frac{\partial}{\partial x^i} \right)_p = 0$, then $\sum \xi^i(p) \left(\frac{\partial}{\partial x^i} \right)_p x^k = 0$, where $x^k \in F(p)$. Then $\xi^k(p) = 0$, by (2.19). Proceeding in this manner, we can say that

$$\xi^1(p) = \xi^2(p) = \cdots = \xi^n(p) = 0.$$

Thus, the set $\{ \frac{\partial}{\partial x^i} : i = 1, 2, 3, \ldots, n \}$ is linearly independent. We can now state the following.

Theorem 2.4 *If $(x^1, x^2, \ldots, x^n)$ is a local coordinate system in a neighbourhood U of a point p in an n-dimensional manifold M, the basis of the tangent space $T_p(M)$ is given by $\{ \frac{\partial}{\partial x^i} : i = 1, 2, 3, \ldots, n \}$ and every $X_p \in T_p(M)$ can be expressed uniquely by (2.22).*

Problem 2.27 Let $X = (2, 3, 0) \in \mathbb{R}^3$. Find $X_p f$ for a fixed point $p = (-2, \pi, 1)$ where $f = x^1 x^3 \cos x^2$.

Solution: 27 Comparing with (2.22), we see that

$$\xi^1(p) = 2, \ \xi^2(p) = 3, \ \xi^3(p) = 0.$$

Further,

$$\left(\frac{\partial f}{\partial x^1} \right)_p = (x^3 \cos x^2)_p = -1;$$

$$\left(\frac{\partial f}{\partial x^2} \right)_p = (-x^1 x^3 \sin x^3)_p = 0;$$

$$\left(\frac{\partial f}{\partial x^3} \right)_p = 2.$$

Thus $X_p f = -2$.

Problem 2.28 Let $X = 2x \dfrac{\partial}{\partial x} - 2y \dfrac{\partial}{\partial y}$ be a vector in $\mathbb{R}^2$. Find $X_p f$ where $f = 2x + y^3$, $p = (x, y)$.

Solution: 28 As done in the previous problem,

$$\xi^1(p) = 2x, \quad \xi^2(p) = -2y, \quad \frac{\partial f}{\partial x}\Big|_p = 2, \quad \frac{\partial f}{\partial y}\Big|_p = 3y^2.$$

Thus $X_p f = 4x - 6y^3$.

Problem 2.29 Let $X = \dfrac{\partial}{\partial x} + \dfrac{\partial}{\partial z}$ be a vector in $\mathbb{R}^2$. Find $X_p f$ for a fixed point $p = (1, 1, 0)$ where $f = xz \cos y$.

Solution: 29 In this case

$$X_p f = 1(z \cos y)\big|_p + 0(-xz \sin y)\big|_p + 1(x \cos y)\big|_p = \cos 1.$$

Problem 2.30 If C is a constant function on a manifold M and X is a tangent vector to some curve σ on M, show that $X \cdot C = 0$.

Solution: 30 Here by (2.16), we have $X(1) = 0$, i.e. $X \cdot 1 = 0$. Again by virtue of linearity, $X \cdot C = X(C \cdot 1) = C(X \cdot 1)$, as C is a constant function. Hence $X \cdot C = 0$.

Problem 2.31 Let $f = ((x^1)^3 - 2)x^3 + (x^2 x^3 - 1)x^1$. Find $X_p f$, $p = (x^1, x^2, x^3)$.

Solution: 31 Here

$$X_p f = \xi^1(p)\{3(x^1)^2 x^3 + (x^2 x^3 - 1)\}_p + \xi^2(p)(x^1 x^3)_p + \xi^3(p)\{(x^1)^3 - 2 + x^1 x^2\}_p, \text{ by (2.21)}.$$

Exercises

Exercise 2.11 *Find the tangent vector*

(i) *to the curve* $\sigma \in \mathbb{R}^n$ *where* $\sigma^i = a^i + b^i t$, $a^i, b^i \in \mathbb{R}$ *for every* i.

(ii) *to the curve* $\sigma(t) = (t^2, t^3)$ *on* $\mathbb{R}^2$.

(iii) *to the curve* $\sigma_p(t) = \begin{pmatrix} \cos 2t & -\sin 2t \\ \sin 2t & \cos 2t \end{pmatrix} \begin{pmatrix} x \\ y \end{pmatrix}$ *at* $t = 0$.

Exercise 2.12 (i) *Consider the curve* $\gamma(t) = (\cos t, \sin t) \in \mathbb{R}^2, t \in (0, \pi)$. *Find the vector* X *tangent to* γ *at* $\frac{\pi}{4}$. *Calculate* Xf *where* $f : \mathbb{R}^2 \to \mathbb{R}$ *is defined by* $f = 2x + y^3$.

(ii) *Consider the curve* ψ *in* $\mathbb{R}^2$ *defined by* $x = \sin t, y = \cos t, t \in (-\pi, \pi)$ *and the map* $f : \mathbb{R}^2 \to \mathbb{R}$ *defined by* $f(x, y) = x^3 y$. *Find the vector* X *tangent to* ψ *at* $t = \frac{\pi}{2}$ *and compute* Xf.

Exercise 2.13 *Let* $X = (2, -3, 4) \in \mathbb{R}^3$. *For a fixed point* $p = (2, 5, 7)$, *compute* $X_p f$ *where*

(i) $f : \mathbb{R}^3 \to \mathbb{R}$ *is defined by* $f = x^3 y$.

(ii) $f : \mathbb{R}^3 \to \mathbb{R}$ *is defined by* $f = z^7$.

(iii) $f : \mathbb{R}^3 \to \mathbb{R}$ *is defined by* $f = e^x \cos z$.

Answers

2.11 (i) $b^1 \dfrac{\partial}{\partial x^1} + b^2 \dfrac{\partial}{\partial x^2} + \cdots + b^n \dfrac{\partial}{\partial x^n}$ (ii) $2t \dfrac{\partial}{\partial x} + 3t^2 \dfrac{\partial}{\partial y}$ or $2t \dfrac{\partial}{\partial x^1} + 3t^2 \dfrac{\partial}{\partial x^2}$.

(iii) $-2y \dfrac{\partial}{\partial x} + 2x \dfrac{\partial}{\partial y}$.

2.12 (i) $-\dfrac{1}{\sqrt{2}} \dfrac{\partial}{\partial x} + \dfrac{1}{\sqrt{2}} \dfrac{\partial}{\partial x}$; $-\dfrac{1}{2\sqrt{2}}$ (ii) $-\dfrac{\partial}{\partial y}$; -1.

2.13 (i) 96 (ii) $4 \cdot 7^7$ (iii) $2e^2(\cos 7 - 2\sin 7)$.

2.9   Inverse Function Theorem for Smooth Manifold

In continuation with the Inverse Function Theorem for $\mathbb{R}^n$, stated and proved in Chap. 1, the following theorem deals with the study of Inverse Function Theorem for arbitrary smooth manifolds.

Theorem 2.5 *Let M and N be n-dimensional smooth manifolds and $F : M \to N$ be a smooth map. Let $p \in M$. If $F_* : T_p(M) \to T_{F(p)}(N)$ is invertible (i.e. $1 - 1$ and onto) at p, then $\exists$ an open neighbourhood U of p, and an open neighbourhood V of $F(p)$ such that $F : U \to V$ is a diffeomorphism.*

Proof Since M is an n-dimensional smooth manifold and $p \in M$, therefore there exists an admissible coordinate chart $(\tilde{U}, \phi)$ of M such that $p \in \tilde{U}$. For every $q \in \tilde{U}$, let $\left(\dfrac{\partial}{\partial x^1} \Big|_q, \dfrac{\partial}{\partial x^2} \Big|_q, \ldots, \dfrac{\partial}{\partial x^n} \Big|_q \right)$ be a coordinate basis of $T_q(M)$ corresponding to $(\tilde{U}, \phi)$. Again, since N is an n-dimensional smooth manifold and $F(p) \in N$, therefore there exists an admissible coordinate chart $(\tilde{V}, \psi)$ of M such that $F(p) \in \tilde{V}$ and $(\psi \circ F)(p) = 0$. For every $r \in \tilde{V}$, suppose $\left(\dfrac{\partial}{\partial y^1} \Big|_r, \dfrac{\partial}{\partial y^2} \Big|_r, \ldots, \dfrac{\partial}{\partial y^n} \Big|_r \right)$ is a coordinate basis of $T_r(N)$ corresponding to $(\tilde{V}, \psi)$. Here, the matrix representation of the linear map f_* at the point p with respect to some basis, denoted by (f_*), is the $n \times n$ order matrix. Since the linear map f_* is invertible, therefore $\det(f_*) \neq 0$. Furthermore, the map $\psi \circ F \circ \phi^{-1} : \phi(\tilde{U} \cap F^{-1}(\tilde{V})) \to \psi(\tilde{V})$ is smooth. Also, $(\psi \circ F \circ \phi^{-1})(\phi(p)) = \psi(F(p)) = 0$. Moreover, it is clear that $\phi(\tilde{U} \cap F^{-1}(\tilde{V}))$ is an open neighbourhood of $\phi(p) \in \mathbb{R}^n$ and $\psi(\tilde{V})$ is an open neighbourhood of $0 \in \mathbb{R}^n$. Now, by virtue of Inverse Function Theorem for $\mathbb{R}^n$, $\exists$ an open neighbourhood $\bar{U}$ of $\phi(p)$ satisfying

- $\bar{U} \subset \phi(\tilde{U} \cap F^{-1}(\tilde{V}))$;
- $(\psi \circ F \circ \phi^{-1})(\bar{U})$ is an open neighbourhood of 0;
- $\psi \circ F \circ \phi^{-1}$ has a smooth inverse on $(\psi \circ F \circ \phi^{-1})(\bar{U})$.

Set $U = \phi^{-1}(\bar{U})$ and $V = (F \circ \phi^{-1})(\bar{U})$. Our claim is that U is an open neighbourhood of $p \in M$. Since $\bar{U}$ is open and contained in $\phi(\tilde{U} \cap F^{-1}(\tilde{V}))(\subset \phi(\tilde{U}))$, and $\phi(\tilde{U})$ is open, $\bar{U}$ is open in $\phi(\tilde{U})$, hence $\phi^{-1}(\bar{U})$ is open in $\tilde{U}$. Moreover, $\phi^{-1}(\bar{U}) = U(say)$ is open in M. Since $\phi(p) \in \bar{U}$, therefore $p \in U$.

Since $\psi \circ F \circ \phi^{-1}$ has a smooth inverse, therefore $\psi \circ F \circ \phi^{-1}$ is continuous. Also $\phi \circ F^{-1}$ is continuous, as ψ is so. Furthermore, $(F \circ \phi^{-1})(\bar{U}) = V$ is open. Since $\phi(p) \in \bar{U}$, therefore $F(p) \in V$. This shows V is an open neighbourhood of $F(p)$.

Since $\psi \circ F \circ \phi^{-1}$ has an inverse, therefore $\psi \circ F \circ \phi^{-1}$ is one-to-one and onto. Thus, the composite map $\psi^{-1}(\psi \circ F \circ \phi^{-1}) \circ \phi = F$ is also one-to-one and onto. Since $\psi \circ F \circ \phi^{-1}$ has a smooth inverse, $\phi \circ F^{-1} \circ \psi^{-1}$ is smooth. This gives $F^{-1} : V \to U$ is smooth. This completes the proof.

Let M and N be n-dimensional smooth manifolds and $F : M \to N$ be a smooth map. Let $p \in M$. If $\exists$ an open neighbourhood U of p, such that the neighbourhood

$F(U)$ of $F(p)$ is open in N and $F : U \rightarrow F(U)$ is a diffeomorphism, then we say that F is a local diffeomorphism.

Problem 2.32 Let M and N be n-dimensional smooth manifolds and $F : M \rightarrow N$ be a local diffeomorphism. Prove that F is an open map.

Solution: 32 Let $p \in U$. Since F is a local diffeomorphism and $p \in M$, $\exists$ an open neighbourhood V of p such that $F(V)$ is open in N, and the map $F|_V : V \rightarrow F(V)$ is a diffeomorphism. Since U, V are open neighbourhoods of p, $U \cap V$ is also an open neighbourhood of p, in V. Since $F|_V : V \rightarrow F(V)$ is a diffeomorphism, therefore it is a homeomorphism. This implies $F|_V(U \cap V)$ is open in $F(V)$, as $U \cap V$ is open in V. Moreover, $F(V)$ being open in N, therefore $F|_V(U \cap V)$ is open in N. Since $p \in U \cap V$, $F(p) \in F(U \cap V)$. Thus, $F(U \cap V) \subset F(V)$ and $F(U \cap V)$ is an open neighbourhood of $F(p)$.

Remark 2.20 Let M, N, P be smooth manifolds. Let $F : M \rightarrow N$ and $G : N \rightarrow P$ be local diffeomorphisms. The composite map $G \circ F : M \rightarrow P$ is a local diffeomorphism.

Let M and N be respectively n- and m-dimensional smooth manifolds. If $F : M \rightarrow N$ is continuous, and for every $p \in M$, $\exists$ an open neighbourhood U of p such that $F(U)$ is an open neighbourhood of $F(p)$ and the map $F|_U : U \rightarrow F(U)$ is a homeomorphism, then $F : M \rightarrow N$ is said to be a local homeomorphism.

Remark 2.21 If $F : M \rightarrow N$ is a local diffeomorphism, then it is a local homeomorphism.

Remark 2.22 Let M, N be smooth manifolds. Let U be a non-empty open subset of M. Let $F : M \rightarrow N$ be a local diffeomorphism. Then $F|_U : U \rightarrow N$ is a local diffeomorphism.

Note that diffeomorphism implies local diffeomorphism, but the converse is not always true in general.

Example 2.8 Consider the map $f : \mathbb{R}^2 \rightarrow \mathbb{R}^2$ defined by $f(x, y) = (e^x \cos y,$ $e^x \sin y)$. Here $f'(x, y) = \begin{pmatrix} e^x \cos y & -e^x \sin y \\ e^x \sin y & e^x \cos y \end{pmatrix}$ and the Jacobian of the matrix is non-zero, which shows $f'(x, y)$ is invertible. But f is not one-to-one, since it is of period 2π. So f is a local diffeomorphism but not a diffeomorphism.

Problem 2.33 Let M and N be respectively n- and m-dimensional smooth manifolds and $F : M \rightarrow N$ be $1 - 1$ and onto. Suppose F is a local diffeomorphism. Then prove that F is a diffeomorphism.

Solution: 33 Since $F : M \rightarrow N$ is a local diffeomorphism, it is a local homeomorphism and hence continuous. Also F is an open map. Since $F : M \rightarrow N$ is $1 - 1$ and onto, F is continuous and F is an open map, $F : M \rightarrow N$ is a homeomorphism. It remains to show that $F : M \rightarrow N$ is smooth and its inverse is also so.

For any $p \in M$, we have to find an admissible coordinate chart (U, ϕ) of M with $p \in U$, and an admissible coordinate chart (V, ψ) of N with $F(p) \in V$ such that $\psi \circ F \circ \phi^{-1} : \phi(U \cap F^{-1}(V)) \to \psi(V)$ is smooth.

Since $p \in M$, and $F : M \to N$ is a local diffeomorphism, $\exists$ an open neighbourhood $\tilde{U}$ of p such that $F(\tilde{U})$ is an open neighbourhood of $F(p)$ in N, and the map $F|_{\tilde{U}} : \tilde{U} \to F(\tilde{U})$ is a diffeomorphism. Furthermore, $\exists$ an admissible coordinate chart $(\hat{U}, \hat{\phi})$ of M with $p \in \hat{U}$. Also, $\exists$ an admissible coordinate chart $(\hat{V}, \hat{\psi})$ of N with $F(p) \in \hat{V}$. Moreover, $\tilde{U} \cap \hat{U}$ is an open neighbourhood of p. So $((\tilde{U} \cap \hat{U}), \hat{\phi}|_{\tilde{U} \cap \hat{U}})$ forms an admissible chart of M satisfying $p \in \tilde{U} \cap \hat{U}$. Since $F|_{\tilde{U}} : \tilde{U} \to F(\tilde{U})$ is a diffeomorphism, the map $F|_{\tilde{U} \cap \hat{U}} : \tilde{U} \cap \hat{U} \to F(\tilde{U} \cap \hat{U})$ is a diffeomorphism. Since $F : M \to N$ is a homeomorphism, and $\tilde{U} \cap \hat{U}$ is an open neighbourhood of p, $F(\tilde{U} \cap \hat{U})$ is an open neighbourhood of $F(p)$.

Since $(\hat{V}, \hat{\psi})$ is an admissible coordinate chart of N with $\hat{V}$ being an open neighbourhood of $F(p)$, and $F(\tilde{U} \cap \hat{U})$ is an open neighbourhood of $F(p)$, $\hat{V} \cap F(\tilde{U} \cap \hat{U})$ is an open neighbourhood of $F(p)$. Hence, $(\hat{V} \cap F(\tilde{U} \cap \hat{U}), \hat{\psi}|_{\hat{V} \cap F(\tilde{U} \cap \hat{U})})$ forms an admissible coordinate chart of N with $F(p) \in \hat{V} \cap F(\tilde{U} \cap \hat{U})$.

Since the map $F|_{\tilde{U}} : \tilde{U} \to F(\tilde{U})$ is a diffeomorphism, it is smooth. As $((\tilde{U} \cap \hat{U}), \hat{\phi}|_{\tilde{U} \cap \hat{U}})$ forms an admissible chart of M satisfying $p \in \tilde{U} \cap \hat{U}$ and $(\hat{V} \cap F(\tilde{U} \cap \hat{U}), \hat{\psi}|_{\hat{V} \cap F(\tilde{U} \cap \hat{U})})$ forms an admissible coordinate chart of N with $F(p) \in \hat{V} \cap F(\tilde{U} \cap \hat{U})$, therefore

$$(\hat{\psi}|_{\hat{V} \cap F(\tilde{U} \cap \hat{U})}) \circ F|_{\tilde{U}} \circ (\hat{\phi}|_{\tilde{U} \cap \hat{U}})^{-1}$$

is smooth. This proves $F : M \to N$ is smooth.

Finally, since $F : M \to N$ is $1 - 1$ and onto, therefore $F^{-1} : N \to M$ exists and is $1 - 1$ and onto. Also $F^{-1} : N \to M$ is a local diffeomorphism, therefore proceeding as above $F^{-1} : N \to M$ is smooth. This completes the solution.

Problem 2.34 Let M and N be respectively n- and m-dimensional smooth manifolds and $F : M \to N$ be a local diffeomorphism. Then F is a smooth immersion and smooth submersion.

Solution: 34 We wish to show that, for every $p \in M$, rank $F = n = m$. Since $p \in M$ and F is a local diffeomorphism, $\exists$ an open neighbourhood U of p such that $F(U)$ is open in N, and the map $F|_U : U \to F(U)$ is a diffeomorphism. Since $U(\neq \phi)$ is an open subset of M, and M is an n-dimensional smooth manifold, U is an n-dimensional smooth manifold, so $\dim T_p(U) = \dim U = n$. Similarly, $\dim T_{F(p)}(F(U)) = \dim F(U) = m$. Since $F|_U : U \to F(U)$ is a diffeomorphism, $p \in U$, the linear map $f_* : T_p(U) \to T_{F(p)}(F(U))$ at p is an isomorphism (refer to Exercise 2.55), which implies rank $F = \dim T_p(U) = n = m = \dim T_{F(p)}(F(U))$ at p.

Problem 2.35 Let M and N be respectively n- and m-dimensional smooth manifolds and $F : M \to N$ be a smooth immersion and smooth submersion. Then F is a local diffeomorphism.

Solution: 35 Let $p \in M$. Our claim is to find an open neighbourhood U of $p \in M$ such that $F(U)$ is open in N, and the map $F|_U : U \to F(U)$ is a diffeomorphism.

Since F is a smooth immersion and smooth submersion, the linear map $F_* : T_p(M) \to T_{F(p)}(N)$ is $1-1$ and onto at $p \in M$, therefore dim $T_p(M) = $ dim $T_{F(p)}(M)$, *i.e.* $n = m$. By Inverse Function Theorem for manifolds, $\exists$ an open neighbourhood V of $F(p)$ with $F|_U : U \to V$ being a diffeomorphism. Hence $F(U) = (F|_U)(U) = V$. Moreover, $F(U)$ being open, $F|_U : U \to F(U)$ is a diffeomorphism.

Exercises

Exercise 2.14 *Let M and N be n-dimensional smooth manifolds and $F : M \to N$ be a smooth immersion. Then F is a local diffeomorphism.*

Exercise 2.15 *Let M and N be n-dimensional smooth manifolds and $F : M \to N$ be a smooth submersion. Then F is a local diffeomorphism.*

2.10 Vector Field

In classical notation, if to each point p of $\mathbb{R}^3$ or in a domain U of $\mathbb{R}^3$, a vector $\alpha : p \to \alpha(p)$ is specified, then we say that a vector field is given on $\mathbb{R}^3$ or in a domain of $\mathbb{R}^3$. In the same manner, we will introduce a vector field in a manifold M.

A vector field X on M is a correspondence that associates with each point p of M, a vector $X_p \in T_p(M)$. In fact, if $f \in F(M)$ then Xf is defined to be a real-valued function on M, as follows:

$$(Xf)(p) = X_p f. \tag{2.23}$$

A vector field X is called differentiable if Xf is so for every $f \in F(M)$. From (2.22), a vector field X can be expressed as

$$X = \sum_i \xi^i \frac{\partial}{\partial x^i}. \tag{2.24}$$

Let $\chi(M)$ denote the set of all differentiable vector fields on M. We define

$$\begin{cases} (X + Y)f = Xf + Yf \\ (\lambda X)f = \lambda(Xf), \ \forall \ X, Y \in \chi(M), \lambda \in \mathbb{R}. \end{cases} \tag{2.25}$$

It can be shown that $\chi(M)$ is a vector space over $\mathbb{R}$. We also define fX to be a vector field on M as follows:

$$(fX)(p) = f(p)X_p, \ \forall \ p \in M. \tag{2.26}$$

Let us define a mapping $[\ ,\ ] : F(M) \to F(M)$ as

$$[X, Y]f = X(Yf) - Y(Xf). \tag{2.27}$$

Such a bracket is also known as **Lie Bracket** of X and Y.

Exercises

Exercise 2.16 *Show that for every X, Y, Z in $\chi(M)$ and f, g in $F(M)$*
(a) $[X, Y] \in \chi(M)$
(b) $[bX, Y] = b[X, Y] = [X, bY], \ \forall \ b \in \mathbb{R}$
(c) $[X + Y, Z] = [X, Z] + [Y, Z]$
(d) $[X, Y + Z] = [X, Y] + [X, Z]$
(e) $[X, [Y, Z]] + [Y, [Z, X]] + [Z, [X, Y]] = \theta$: Jacobi Identity
(f) $[X, X] = \theta$
(g) $[X, Y] = -[Y, X]$.

Hints
2.16 (a). Show that $[X, Y]$ satisfies Linearity and Leibnitz Product rule.

Remark

Remark 2.23 $\chi(M)$ with the product rule given by (2.27) is an algebra, also called **Lie Algebra**.

Problem 2.36 Using $[X, X] = \theta$, show that $[X, Y] = -[Y, X]$.

Solution: 36 For every $X, Y \in \chi(M), X + Y \in \chi(M)$. Using the hypothesis, we have

$$[X + Y, X + Y] = \theta$$
$$i.e. \ \ [X, Y] + [Y, X] = \theta,$$
$$\therefore \ [X, Y] = -[Y, X].$$

Problem 2.37 Prove that $[X, fY] = f[X, Y] + (Xf)Y$.

Solution: 37 Note that

$$((fX)h)(p) = (fX)_p h, \ \forall \ h \in F(M) \text{ by } (2.23)$$
$$= f(p)X_p h, \ \text{by } (2.26).$$
$$\text{Also, } (f(Xh))(p) = f(p)(Xh)(p), \ \text{by } (1.8)$$
$$= f(p)X_p h, \ \text{by } (2.23).$$

Thus

$$(fX)h = f(Xh), \ \forall \ p \in M. \tag{2.28}$$

Again (2.27) yields

$$[X, fY] = X\{(fY)h\} - (fY)(Xh)$$
$$= X\{f(Yh)\} - f\{Y(Xh)\}, \text{ by (2.28)}$$
$$= (Xf)(Yh) + f\{X(Yh)\} - f\{Y(Xh)\}, \text{ by (2.17)}$$
$$= \{(Xf)Y\}h + f\{[X, Y]h\}, \text{ by (2.27) and (2.28)}$$
$$= \{(Xf)Y\}h + \{f[X, Y]\}h, \text{ by (2.28)}$$

$i.e., \ [X, fY] = f[X, Y] + (Xf)Y, \ \forall h.$

Problem 2.38 If $X = \dfrac{\partial}{\partial x}, Y = \dfrac{\partial}{\partial y} + e^x \dfrac{\partial}{\partial z}$, compute $[X, Y]_{(0,1,0)}$.

Solution: 38 Note that

$$[X, Y]f = \frac{\partial}{\partial x}\{\frac{\partial f}{\partial y} + e^x \frac{\partial f}{\partial z}\} - (\frac{\partial}{\partial y} + e^x \frac{\partial}{\partial z})\frac{\partial f}{\partial x}$$

$$= e^x \frac{\partial f}{\partial z}$$

$$\therefore, \ [X, Y] = e^x \frac{\partial}{\partial z}, \ \forall f$$

Hence, $[X, Y]_{(0,1,0)} = \dfrac{\partial}{\partial z}\Big|_{(0,1,0)}$.

Problem 2.39 Find the general expression for $Z \in \chi(\mathbb{R}^2)$ where

$$\left[\frac{\partial}{\partial x_1}, Z\right] = Z \quad \text{and} \quad \left[\frac{\partial}{\partial x_2}, Z\right] = Z.$$

Solution: 39 Let us assume that

$$Z = \lambda(x_1, x_2)\frac{\partial}{\partial x_1} + \mu(x_1, x_2)\frac{\partial}{\partial x_2}, \ \lambda, \mu \in F(\mathbb{R}^2).$$

Substituting the expression of Z and using (2.27), one gets after a few steps

$$\left[\frac{\partial}{\partial x_1}, Z\right] = \frac{\partial \lambda(x_1, x_2)}{\partial x_1}\frac{\partial}{\partial x_1} + \frac{\partial \mu(x_1, x_2)}{\partial x_1}\frac{\partial}{\partial x_2}.$$

Similarly,

$$\left[\frac{\partial}{\partial x_2}, Z\right] = \frac{\partial \lambda(x_1, x_2)}{\partial x_2}\frac{\partial}{\partial x_1} + \frac{\partial \mu(x_1, x_2)}{\partial x_2}\frac{\partial}{\partial x_2}.$$

From the given condition,

$$\left[\frac{\partial}{\partial x_1}, Z\right] = Z \quad \text{and} \quad \left[\frac{\partial}{\partial x_2}, Z\right] = Z.$$

Thus on comparing, we have

$$\frac{\partial \lambda(x_1, x_2)}{\partial x_1} = \lambda(x_1, x_2); \quad \frac{\partial \mu(x_1, x_2)}{\partial x_1} = \mu(x_1, x_2); \quad \frac{\partial \lambda(x_1, x_2)}{\partial x_2} = \lambda(x_1, x_2); \quad \frac{\partial \mu(x_1, x_2)}{\partial x_2} = \mu(x_1, x_2).$$

From the first equation, we see that

$$\lambda(x_1, x_2) = A f(x_2) e^{x_1}, \quad A \text{ being constant.}$$

Substituting in the third equation, we get

$$\frac{\partial}{\partial x_2} \{A f(x_2) e^{x_1}\} = A f(x_2) e^{x_1}$$

$$A f'(x_2) e^{x_1} = A f(x_2) e^{x_1}$$

$$\therefore, \ f(x_2) = C e^{x_2}, \ C \text{ is constant.}$$

Thus $\lambda(x_1, x_2) = A C e^{x_2} e^{x_1} = D e^{x_1 + x_2}$, say $D = AC$ being constant.

By similar computation, from the second and fourth equations, it can be found that $\mu(x_1, x_2) = B g(x_2) e^{x_1}$ and after a brief calculation, $\mu(x_1, x_2) = E e^{x_1 + x_2}$, E being a constant. Thus,

$$Z = D e^{x_1 + x_2} \frac{\partial}{\partial x_1} + E e^{x_1 + x_2} \frac{\partial}{\partial x_2}.$$

Problem 2.40 Write in cylindrical coordinates, the vector field on $\mathbb{R}^3$ defined by

$$X = \frac{\partial}{\partial x} + \frac{\partial}{\partial y} + \frac{\partial}{\partial z}.$$

Solution: 40 If (ρ, θ, z) is the cylindrical coordinate, then the Cartesian coordinate (x, y, z) is given by

$$x = \rho \cos \theta, \ y = \rho \sin \theta, \ z = z.$$

Therefore, $|J| = \rho$. Let us write

$$X = \xi^1(\rho, \theta, z) \frac{\partial}{\partial \rho} + \xi^2(\rho, \theta, z) \frac{\partial}{\partial \theta} + \xi^3(\rho, \theta, z) \frac{\partial}{\partial \theta}.$$

Then $\ J \begin{pmatrix} \xi^1 \\ \xi^2 \\ \xi^3 \end{pmatrix} = \begin{pmatrix} 1 \\ 1 \\ 1 \end{pmatrix}$, i.e., $\xi^1 \cos \theta - \xi^2 \rho \sin \theta = 1$; $\xi^1 \sin \theta + \xi^2 \rho \cos \theta = 1$; $\xi^3 = 1$. After a few steps, one gets from above

$$\xi^1 = \cos\theta + \sin\theta$$

$$\xi^2 = \frac{1}{\rho}(\cos\theta - \sin\theta).$$

Thus in cylindrical coordinate, we have

$$X = (\cos\theta + \sin\theta)\frac{\partial}{\partial\rho} + \frac{1}{\rho}(\cos\theta - \sin\theta)\frac{\partial}{\partial\theta} + \frac{\partial}{\partial z}.$$

Problem 2.41 If $X = (x - y)\frac{\partial}{\partial x} - \frac{\partial}{\partial y}$, $Y = x^2\frac{\partial}{\partial x} + y\frac{\partial}{\partial y}$ are vector fields on $\mathbb{R}^2$, show that X, Y are linearly independent differentiable vector fields on $\mathbb{R}^2$, if $x^2 - y^2 + xy \neq 0$. Further, if $Z = (x^2 - y^2)\frac{\partial}{\partial x} + (x^2 + y^2)\frac{\partial}{\partial y}$ is any vector of $\mathbb{R}^2$, express $Z = fX + gY$, where $f, g \in F(\mathbb{R}^2)$.

Solution: 41 Clearly, X, Y are differentiable, as $x - y, x^2, y$ are also so. If X, Y are linearly independent, then for $\lambda X + \mu Y = \theta \Rightarrow \lambda = \mu = 0$, $\forall\, \lambda, \mu \in \mathbb{R}$. Again $\lambda X + \mu Y = \theta$ gives

$$\{\lambda(x - y) + \mu x^2\}\frac{\partial}{\partial x} + (-\lambda + \mu y)\frac{\partial}{\partial y} = \theta.$$

As $\left\{\frac{\partial}{\partial x}, \frac{\partial}{\partial y}\right\}$ is a basis of $T_{(x,y)}(\mathbb{R}^2)$, we must have

$$\lambda(x - y) + \mu x^2 = 0 = -\lambda + \mu y.$$

Therefore, $\lambda = \mu y$ and $\mu(x^2 - y^2 + xy) = 0$. Thus X, Y are linearly independent if $x^2 - y^2 + xy \neq 0$. Writing $Z = fX + gY$, we find on comparing

$$x^2 - y^2 = f(x - y) + gx^2 \quad \text{and} \quad x^2 + y^2 = -f + gy.$$

One gets, after a brief calculation,

$$f = \frac{x^2 y - y^3 - x^4 - x^2 y^2}{x^2 + xy - y^2}, \quad g = \frac{(x - y)(x + y + x^2 + y^2)}{x^2 + xy - y^2}.$$

Exercises

Exercise 2.17 *Show that*
(a) $[fX, Y] = f[X, Y] - (Yf)X$
(b) $[fX, gY] = fg[X, Y] + \{f(Xg)\}Y - \{g(Yf)\}X$, *where* $X, Y \in \chi(M)$ *and* $f, g \in F(M)$.

Exercise 2.18 *In terms of a local coordinate system* $(x^1, x^2, \dots, x^n)$ *of a point in a neighbourhood of a differential manifold* M, *show that*

(a) $\left[\dfrac{\partial}{\partial x^i}, \dfrac{\partial}{\partial x^j}\right] = \theta,\ i, j = 1, 2, 3, \ldots, n.$

(b) $[X, Y] = \displaystyle\sum_{i,j}\left(\xi^i \dfrac{\partial \eta^j}{\partial x^i} - \eta^i \dfrac{\partial \xi^j}{\partial x^i}\right)\dfrac{\partial}{\partial x^j}$ where $X = \displaystyle\sum_{i=1}^{n} \xi^i \dfrac{\partial}{\partial x^i},\ Y = \displaystyle\sum_{j=1}^{n} \eta^j \dfrac{\partial}{\partial x^j},$

each $\xi^i, \eta^j \in F(M).$

Exercise 2.19 *Compute* $\left[y\dfrac{\partial}{\partial x} - x\dfrac{\partial}{\partial y}, \dfrac{\partial}{\partial x}\right].$

Exercise 2.20 *Let* $X = x\dfrac{\partial}{\partial x} + y\dfrac{\partial}{\partial y}.$ *Compute* Xf *where*

(i) $f : \mathbb{R}^3 \to \mathbb{R}$ *is defined by* $f(x, y, z) = x^2 - y^2 - z^2$
(ii) $f : \mathbb{R}^2 \to \mathbb{R}$ *is defined by* $f(x, y) = xy^7$
(iii) $f : \mathbb{R}^3 \to \mathbb{R}$ *is defined by* $f(x, y, z) = e^x \cos y.$

Exercise 2.21 *(A). Compute* $[X, Y]$*; (B). Compute* $[X, Y]_{(1,0)}$ *where*

(i) $X = \dfrac{\partial}{\partial x}, Y = e^x \dfrac{\partial}{\partial y} + \dfrac{\partial}{\partial x}$

(ii) $X = x^2 \dfrac{\partial}{\partial x}, Y = x\dfrac{\partial}{\partial y}$

(iii) $X = x^2 \dfrac{\partial}{\partial x} + y^2 \dfrac{\partial}{\partial y}, Y = (y + 1)\dfrac{\partial}{\partial x}.$

Exercise 2.22 *(A). Compute* $[X, Y]$*; (B). Compute* $[X, Y]_{(1,1,1)}$ *where*

(i) $X = \dfrac{\partial}{\partial x}, Y = e^x \dfrac{\partial}{\partial y} + \dfrac{\partial}{\partial z}$

(ii) $X = y\dfrac{\partial}{\partial x} + x\dfrac{\partial}{\partial z}, Y = y\dfrac{\partial}{\partial y}.$

Exercise 2.23 *Compute (A).* $(fX)_{(1,1,1)}$ *and (B).* $(Xf)_{(1,1,1)}$ *where* $f : \mathbb{R}^3 \to \mathbb{R}$ *is defined*

by $f(x, y, z) = x^2 y^2$ *and (i)* $X = y\dfrac{\partial}{\partial x} + x\dfrac{\partial}{\partial z},$ *(ii)* $X = e^x \dfrac{\partial}{\partial y} + \dfrac{\partial}{\partial z}$ *and (iii)* $X = \dfrac{\partial}{\partial y} + e^z \dfrac{\partial}{\partial z}.$

Answers

2.19. $\dfrac{\partial}{\partial y}.$

2.20. (i). $2(x^2 - y^2)$ (ii) $8xy^7$ (iii) $e^x(x \cos y) - y \sin y.$

2.21. (A) (i) $\dfrac{\partial}{\partial y}$ (ii) $x^2 \dfrac{\partial}{\partial y}$ (iii) $y^2 \dfrac{\partial}{\partial x} - 2x(y + 1)\dfrac{\partial}{\partial x}.$

　　(B) (i) $\dfrac{\partial}{\partial y}\Big|_{(1,0)}$ (ii) $\dfrac{\partial}{\partial y}\Big|_{(1,0)}$ (iii) $-2\dfrac{\partial}{\partial x}\Big|_{(1,0)}.$

2.22 (A) (i) $\dfrac{\partial}{\partial y}$ (ii) $-y\dfrac{\partial}{\partial x}$ (B)(i) $\dfrac{\partial}{\partial y}\big|_{(1,1,1)}$ (ii) $-\dfrac{\partial}{\partial x}\big|_{(1,1,1)}$.

2.23 (A) (i) $\dfrac{\partial}{\partial x}\big|_{(1,1,1)} + \dfrac{\partial}{\partial z}\big|_{(1,1,1)}$ (ii) $e\dfrac{\partial}{\partial y}\big|_{(1,1,1)} + \dfrac{\partial}{\partial z}\big|_{(1,1,1)}$ (iii) $\dfrac{\partial}{\partial y}\big|_{(1,1,1)} + e\dfrac{\partial}{\partial z}\big|_{(1,1,1)}$.

(B) (i) 2 (ii) $2e$ (iii) 2.

2.11 Integral Curve

We are going to state the geometrical interpretation of the vector field in this section.

In the last section, we have shown that a vector field is a rule that gives a tangent vector at every point of the manifold M. Each point of M has its own tangent space. The question now arises—for a given vector field, can we start from one point of M and choose a curve whose tangent vector is always the given vector field? The answer has been given in the affirmative sense.

At $p \in U \subset M$, suppose a vector field $Y \in \chi(M)$ is specified. A curve σ is an **integral curve** of the vector field Y if the range of σ is contained in U and for every $a \le t_0 \le b$ in the domain $[a, b]$ of $\mathbb{R}$ of σ, the tangent vector to σ at $\sigma(t_0) = p$ coincides with Y_p, i.e.

$$Y_p = Y_{\sigma(t_0)}$$
$$i.e.\ Y_p f = Y_{\sigma(t_0)} f, \ \forall \ f \in F(M).$$

Using (2.15) and (2.22), we see that

$$\sum_{i=1}^{n} \xi^i(p)\left(\frac{\partial}{\partial x^i}\right)_p f = \frac{d}{dt}(f \circ \sigma)(t)\bigg|_{t=t_0} = \sum_{i=1}^{n} \frac{dx^i(t)}{dt}\bigg|_{t=t_0} \left(\frac{\partial}{\partial x^i}\right)_p f.$$

Since $\{\left(\dfrac{\partial}{\partial x^i}\right)_p : i = 1, 2, 3, \ldots, n\}$ is a basis of $T_p(M)$, we must have

$$\xi^i(p) = \frac{dx^i(t)}{dt}\bigg|_{t=t_0}$$

$$\text{or } \xi^i(\sigma(t))\bigg|_{t=t_0} = \frac{dx^i(t)}{dt}\bigg|_{t=t_0}$$

$$i.e.\ \xi^i(x^1(t), x^2(t), \ldots, x^n(t))\bigg|_{t=t_0} = \frac{dx^i(t)}{dt}\bigg|_{t=t_0}, \text{ by (2.14).}$$

Hence they are related by

$$\frac{dx^i(t)}{dt} = \xi^i(x^1(t), x^2(t), \ldots, x^n(t)). \tag{2.29}$$

A vector field X on M is said to be **complete** if at every point p of M, the integral curve of X through p can be defined for all $t \in \mathbb{R}$. Otherwise, it is said to be an incomplete vector field.

An integral curve is said to be **maximal** if its domain cannot be extended to a larger interval.

Remark 2.24 Note that the paths of different integral curves can never cross except possibly at a point where $\xi^i = 0$ for all t, because of the uniqueness of the solutions of (2.29). Since some integral curve passes through each point p (it is found by solving (2.29) with initial conditions at p), the integral curves "fill" M.

For instance, if M is a 3-dimensional manifold, then there exists a 2-dimensional family of integral curves for each vector field on M and they cover all of M. Such a manifold-filling set of integral curves is called **congruence**.

Remark 2.25 Let σ_1 and σ_2 be integral curves of a vector field X defined on open intervals I_1 and I_2 respectively, containing 0. If $\sigma_1(0) = \sigma_2(0)$, then $\sigma_1 = \sigma_2$ at each point of $I_1 \cap I_2$.

Problem 2.42 Find the integral curve of the null vector field.

Solution: 42 For a given null vector field on $\mathbb{R}^n$, the required differential equations are

$$\frac{dx^1}{dt} = \frac{dx^2}{dt} = \cdots = \frac{dx^n}{dt} = 0, \tag{2.30}$$

where $\theta = 0\dfrac{\partial}{\partial x^1} + 0\dfrac{\partial}{\partial x^2} + \cdots + 0\dfrac{\partial}{\partial x^n}$.

If for initial condition $t = 0$, we have $x^1 = p^1, x^2 = p^2, \ldots, x^n = p^n$, then we get from (2.30) after integration

$$c^1 = p^1, c^2 = p^2, \ldots, c^n = p^n,$$

where $c^1, c^2, \ldots, c^n$ are integrating constants.

Thus the integral curve, say σ, for the null vector field θ on $\mathbb{R}^n$, is given by

$$\sigma = (p^1, p^2, \ldots, p^n), \quad i.e. \text{ the point itself.}$$

Problem 2.43 Compute the integral curve of the vector field $X = -y\dfrac{\partial}{\partial x} + x\dfrac{\partial}{\partial y}$ on $\mathbb{R}^2$, starting at the point $(1, 0) \in \mathbb{R}^2$.

Solution: 43 The differential equations are

$$\frac{dx}{dt} = -y, \quad \frac{dy}{dt} = x.$$

Thus $\dot{x} = -y$ gives $\ddot{x} = -\dot{y} = -x$ from above. Hence $x = A\cos t + B\sin t$, where A, B are to be determined. Therefore $y = -\dot{x}$ gives $y = A\sin t - B\cos t$. It is given

Fig. 2.18 Integral curves of the vector field $X = -y\frac{\partial}{\partial x} + x\frac{\partial}{\partial y}$

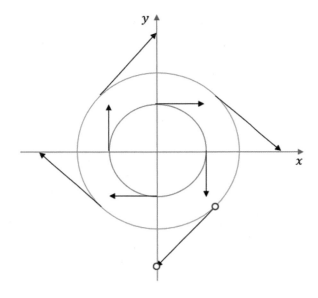

that, for $t = 0$, $x = 1$, $y = 0$. Hence we get from the above

$$A = 1, B = 0 \ i.e. \ x = \cos t, \ y = \sin t.$$

Thus the integral curve σ for the given vector field, starting at $(1, 0)$, is $\sigma = (\cos t, \sin t)$, i.e. the curve is the unit circle.

Remark 2.26 In general, if for $t = 0$, $x = p^1$ and $y = p^2$, then

$$p^1 = A, \quad p^2 = B,$$

i.e. $x(t) = p^1 \cos t + p^2 \sin t$, $y(t) = p^1 \sin t - p^2 \cos t$. Hence, the integral curve σ for the given vector field, starting from $p = (p^1, p^2)$, is

$$\sigma(t) = (p^1 \cos t + p^2 \sin t, p^1 \sin t - p^2 \cos t).$$

It is to be noted that $\sigma(t)$ is defined for all $t \in \mathbb{R}$ and hence the given vector field X is a complete vector field.

In this case,

$$x^2(t) + y^2(t) = (p^1)^2 + (p^2)^2.$$

Thus the integral curves are circles with centre at the origin. The figure is given (Fig. 2.18).

Problem 2.44 Let $X = y\frac{\partial}{\partial x}$, $Y = \frac{x^2}{2}\frac{\partial}{\partial y}$ be two vector fields on $\mathbb{R}^2$. Show that X, Y are complete but $[X, Y]$ is not.

Solution: 44 For $X = y\dfrac{\partial}{\partial x}$, the differential equations are

$$\frac{dx}{dt} = y, \quad \frac{dy}{dt} = 0.$$

After integration, $x = yt + A$, $y = B$, where A, B are integrating constants. For $t = 0$, if $x = x_0$, $y = y_0$, then $A = x_0$, $B = y_0$. Consequently the integral curve, say $\sigma(t)$ for X, through (x_0, y_0) is

$$\sigma(t) = (y_0 t + x_0, y_0),$$

which is defined for all $t \in \mathbb{R}$. Hence X is complete.

Similarly, for $Y = \dfrac{x^2}{2}\dfrac{\partial}{\partial y}$ in $\chi(\mathbb{R}^2)$, the differential equations are

$$\frac{dx}{dt} = 0, \quad \frac{dy}{dt} = \frac{x^2}{2}.$$

Consequently the integral curve, say $\tilde{\sigma}(t)$ for Y, through (x_0, y_0), is given by

$$\tilde{\sigma}(t) = \left(x_0, \frac{1}{2}x_0^2 + y_0\right),$$

which is defined for all $t \in \mathbb{R}$. Hence Y is complete.

Now

$$[X, Y]f = \left[y\frac{\partial}{\partial x}, \frac{x^2}{2}\frac{\partial}{\partial y}\right]f = y\frac{\partial}{\partial x}\left(\frac{x^2}{2}\frac{\partial f}{\partial y}\right) - \frac{x^2}{2}\frac{\partial}{\partial y}\left(y\frac{x^2}{2}\frac{\partial f}{\partial x}\right)$$

$$= yx\frac{\partial f}{\partial y} + \frac{yx^2}{2}\frac{\partial^2 f}{\partial x \partial y} - \frac{x^2}{2}\frac{\partial f}{\partial x} - \frac{x^2 y}{2}\frac{\partial^2 f}{\partial x \partial y}$$

i.e. $[X, Y] = -\dfrac{x^2}{2}\dfrac{\partial}{\partial x} + xy\dfrac{\partial}{\partial y}.$

Thus, the differential equations are

$$\frac{dx}{dt} = -\frac{x^2}{2}, \quad \frac{dy}{dt} = xy. \tag{2.31}$$

Integrating the foregoing equation, one finds

$$\frac{1}{x} = \frac{t}{2} + A, \quad A \text{ being constant.}$$

Thus, $A = \dfrac{1}{x_0}$ for $t = 0$, $x = x_0$ and hence $x = \dfrac{2x_0}{x_0 t + 2}$. From (2.31), one gets

$$\frac{dy}{y} = \frac{2x_0}{x_0 t + 2} dt.$$

On integrating,

$$\log y = 2 \log(x_0 t + 2) + \log B,$$

where $\log B$ is the integrating constant. Therefore $y = (x_0 t + 2)^2$. Hence $y_0 = 4B$, where $y = y_0$ for $t = 0$. Consequently, the integral curve $\gamma(t)$ for $[X, Y]$ is

$$\gamma(t) = \left(\frac{2x_0}{x_0 t + 2}, \frac{y_0}{4} (x_0 t + 2)^2 \right),$$

which is not defined for $t = -\frac{2}{x_0}$. Thus $[X, Y]$ is not complete.

Problem 2.45 Let $X = y \frac{\partial}{\partial x}$, $Y = x \frac{\partial}{\partial y}$ be two vector fields on $\mathbb{R}^2$. Show that X, Y are complete. Is $[X, Y]$ a complete vector field?

Solution: 45 For $X = y \frac{\partial}{\partial x}$, the differential equations are

$$\frac{dx}{dt} = y, \quad \frac{dy}{dt} = 0.$$

After integration, $x = yt + A, y = B$, where A, B are integrating constants. For $t = 0$, if $x = x_0, y = y_0$, then $A = x_0, B = y_0$. Consequently the integral curve, say $\sigma(t)$ for X, through (x_0, y_0) is

$$\sigma(t) = (y_0 t + x_0, y_0),$$

which is defined for all $t \in \mathbb{R}$. Hence X is complete.

Similarly, for $Y = \frac{x}{2} \frac{\partial}{\partial y}$ in $\chi(\mathbb{R}^2)$, the differential equations are

$$\frac{dx}{dt} = 0, \quad \frac{dy}{dt} = x.$$

In a similar manner, we can show that the integral curve, say $\hat{\sigma}(t)$ for Y, through (x_0, y_0) is given by

$$\hat{\sigma}(t) = (x_0, x_0 t + y_0),$$

which is defined for all $t \in \mathbb{R}$. Hence Y is complete.

Again, after a brief calculation we obtain

$$[X, Y] = -x \frac{\partial}{\partial x} + y \frac{\partial}{\partial y}.$$

Hence, the differential equations are

$$\frac{dx}{dt} = -x, \quad \frac{dy}{dt} = y,$$

which on solving, one gets

$$x = -x_0 e^t, \quad y = y_0 e^t, \text{ where } x = x_0, y = y_0 \text{ for } t = 0.$$

Thus, the integral curve $\gamma(t)$ of $[X, Y]$ through (x_0, y_0) is given by $\gamma(t) = (-x_0 e^t, y_0 e^t)$, which is defined for all $t \in \mathbb{R}$. Thus $[X, Y]$ is a complete vector field.

Problem 2.46 Find the integral curve for a given vector field $X = x\dfrac{\partial}{\partial x} + y\dfrac{\partial}{\partial y}$ in $\mathbb{R}^2$. Is X complete? Give the geometrical interpretation of such X.

Solution: 46 The differential equations are $\dfrac{dx}{dt} = x$ and $\dfrac{dy}{dt} = y$. Integrating one gets $\log x = t + A$ and $\log y = t + B$, A, B being integration constants. With initial condition, for $t = 0$, let $x = p^1, y = p^2$. Then

$$x = p^1 e^t, \quad y = p^2 e^t.$$

Hence the integral curve, say σ, for X is given by

$$\sigma = (p^1 e^t, p^2 e^t),$$

which is defined for all $t \in \mathbb{R}$. Thus X is complete. Also, $\dfrac{x}{y} = c$, say where $c = \dfrac{p^1}{p^2}$. Therefore $x = cy$. This represents straight lines passing through the origin of $\mathbb{R}^2$.

Problem 2.47 Let X be the vector field $x^2\dfrac{\partial}{\partial x}$ on the real line $\mathbb{R}$. Find the integral curve of X at 1. Is X complete?

Solution: 47 The differential equation is

$$\frac{dx}{dt} = x^2.$$

Integrating, one gets

$$-\frac{1}{x} = t + A, \quad A \text{ being integration constant.}$$

When $t = 0$, then $x = 1$. Thus $A = -1$. Consequently, $x = \dfrac{1}{1-t}$. Hence the integral curve, say σ of X, is $\sigma = \dfrac{1}{1-t}$ which is not defined for $t = 1$. Thus X is not a complete vector field.

Exercises

Exercise 2.24 *Find the integral curve for the following vector fields. Also check whether the given vector field is complete or not:*

(a) $X = e^{-x}\dfrac{\partial}{\partial x}$ *on* $\mathbb{R}$.

(b) $X = \dfrac{\partial}{\partial x^1} + (x^1)^2\dfrac{\partial}{\partial x^2}$ *on* $\mathbb{R}^2$.

(c) $X = \dfrac{\partial}{\partial y} + e^x\dfrac{\partial}{\partial z}$ *on* $\mathbb{R}^3$.

(d) $X = y\dfrac{\partial}{\partial x} - x\dfrac{\partial}{\partial y}$ *on* $\mathbb{R}^2$.

(e) $X = x^2\dfrac{\partial}{\partial x^1} - (x^2)^3\dfrac{\partial}{\partial x^2}$ *on* $\mathbb{R}^2$.

(f) $X = \dfrac{\partial}{\partial x}$ *where* $X \in \chi(\mathbb{R}^2 - \{0\})$.

Exercise 2.25 *Compute the integral curve of the vector field* $X = \dfrac{\partial}{\partial x} + 2y\dfrac{\partial}{\partial y} + 3\dfrac{\partial}{\partial z}$ *on* $\mathbb{R}^3$ *passing through* (x_0, y_0, z_0) *at* $t = 0$.

Exercise 2.26 *Compute the integral curve of* $X = \dfrac{\partial}{\partial x} + x\dfrac{\partial}{\partial y}$ *on* $\mathbb{R}^2$ *passing through* (a, b) *at* $t = 0$.

Exercise 2.27 *Let* X *be the vector field* $x\dfrac{\partial}{\partial x}$ *on* $\mathbb{R}$. *Find the integral curve of* X *starting at* p.

Exercise 2.28 *Find the integral curve of the vector field* $X = \left(\dfrac{x+y}{r}\right)\dfrac{\partial}{\partial y} - \left(\dfrac{y-x}{r}\right)\dfrac{\partial}{\partial x}$ *on* $\mathbb{R}^2$.

Answers

2.24. (a) $\log(t + e^p)$; No (b) $(p^1 + t, t(p^1 + t)^2, p^2)$; Yes
   (c) $(p^1, t + p^2, te^1 + p^2)$; Yes (d) $(p^1\cos t + p^2\sin t, -p^1\sin t + p^2\cos t)$; Yes

(e) $\left(\dfrac{tp^2}{\sqrt{1 - 2t(p^2)^2}} + p^1, \dfrac{p^2}{\sqrt{1 - 2t(p^2)^2}}\right)$; No (f) $(t + p^1, p^2)$; No

2.25. $(t + x_0, y_0e^{2t}, 3t + z_0)$ 2.26. $(t + a, t(t + a), b)$ 2.27. pe^t
2.28. family of logarithmic spiral, where $r = \sqrt{x^2 + y^2}$.

2.12 Differential of a Mapping

Let M be an n-dimensional and N be an m-dimensional manifold and $f : M \to N$ be a C^∞ map (Fig. 2.19).

Such f induces a map

$$f^* : F(f(p)) \to F(p), \ \ \text{by}$$

$$f^*(g) = g \circ f, \ \forall g \in F(f(p)), p \in M \tag{2.32}$$

and is called the **pull-back** of g by f. It satisfies

$$\begin{cases} f^*(ag + bh) = a(f^*g) + b(f^*h) \\ f^*(gh) \quad = f^*(g)f^*(h), \ \forall h \in F(f(p)), a, b \in \mathbb{R}. \end{cases} \tag{2.33}$$

The map f also induces a mapping

$$f_* : T_p(M) \to T_{f(p)}(N), \ \ \text{such that}$$

$$\{f_*(X_p)\}g = X_p(f^*g) = X_p(g \circ f) \tag{2.34}$$

and is called the **push-forward** of X by f at p, denoted by $f_{*,p}$. Such f_* is also called the **derived linear map** or **differential map of** f **on** $T_p(M)$. We write

$$f_*(X_p) = (f_*X)_{f(p)}. \tag{2.35}$$

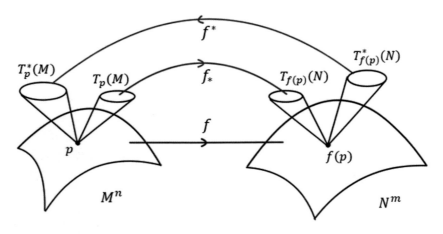

Fig. 2.19 Push-forward mapping

Remark 2.27 These notational conventions f_*, f^* as defined above reflect a similar situation in linear algebra related to linear mappings of vector spaces and their duals, respectively.

Problem 2.48 Prove that f_* is a linear map.

Solution: 48 Note that, for X_p, $Y_p \in T_p(M)$ and $\lambda \in \mathbb{R}$, we have

$$
\begin{aligned}
\{f_*(\lambda X_p + Y_p)\}g = (\lambda X_p + Y_p)(f^*g) &= (\lambda X_p + Y_p)(g \circ f), \text{ by (2.34)} \\
&= \lambda X_p(g \circ f) + Y_p(g \circ f) \\
&= \lambda \{f_*(X_p)\}g + \{f_*(Y_p)\}g, \text{ by (2.34)}.
\end{aligned}
$$

This proves f_* is a linear map.

Problem 2.49 Prove that $f_*(X_p)$ is the derivation at $f(p)$.

Solution: 49 Note that, for all $h, g, h + g \in F(f(p))$, we obtain

$$
\begin{aligned}
\{f_*(X_p)\}(h + g) = X_p(f^*(h + g)) &= X_p((h + g) \circ f) \\
&= X_p(h \circ f) + X_p(g \circ f) \\
&= \{f_*(X_p)\}(h) + \{f_*(X_p)\}(g).
\end{aligned}
$$

Also

$$
\{f_*(X_p)\}(\lambda h) = \lambda X_p(h \circ f) = \lambda \{f_*(X_p)h\}, \ \lambda \in \mathbb{R}.
$$

Thus $f_*(X_p)$ is the derivation at $f(p)$.

Problem 2.50 If I is the identity map in the neighbourhood of a point p in a manifold M, prove that $(I_*)_p$ is the identity map on $T_p(M)$.

Solution: 50 Let I_M denote the identity C^∞ map in the neighbourhood of a point p of M. By (2.34), we obtain

$$
\{(I_*)_p X_p\}g = X_p(g \circ I) = X_p g,
$$

$$
\therefore \ (I_*)_p X_p = X_p, \ \forall \, g.
$$

Thus $(I_*)_p$ is the identity differential of $T_p(M)$.

Problem 2.51 If f is a smooth map from a manifold M into another manifold N and g is a smooth map from N into another manifold L, then

$$
(g \circ f)_* = g_* \circ f_*.
$$

Solution: 51 Note that $g \circ f : M \to L$. Now $f, g, g \circ f$ induce the following linear map:

$$f_* : T_p(M) \to T_{f(p)}(N), \ g_* : T_{f(p)}(N) \to T_{g(f(p))}(L), \ (g \circ f)_* : T_p(M) \to T_{(g \circ f)(p)}(L).$$

Let $h \in F(L)$. Then $h \in F((g \circ f)(p))$. Now

$$\begin{aligned}
((g \circ f)_* X_p)h &= X_p(h \circ (g \circ f)) \\
&= X_p((h \circ g) \circ f) \\
&= (f_*(X_p))(h \circ g) \\
&= \{g_*(f_*(X_p))\}h \\
\therefore \ (g \circ f)_* X_p &= g_*(f_*(X_p)), \ \forall h \\
\text{or } (g \circ f)_* &= g_* \circ f_*, \ \forall X_p \in T_p(M).
\end{aligned}$$

Problem 2.52 Let $f : M \to N$ be a diffeomorphism between two manifolds M and N. Prove that

$$f_*^{-1}(gX) = (g \circ f)f_*^{-1}X, \ \forall g \in F(N).$$

Solution: 52 Given that $f : M \to N$ is a diffeomorphism and hence by definition, $f^{-1} : N \to M$ is C^∞. Thus we can write

$$\begin{aligned}
\{f_*^{-1}(X_{f(p)})\}h &= X_{f(p)}(h \circ f^{-1}), \text{ by } (2.34), \ \forall h \in F(M) \\
\text{or } (f_*^{-1}X)_p h &= X_{f(p)}(h \circ f^{-1}), \text{ by } (2.35).
\end{aligned}$$

Now for all $X \in \chi(N), gX \in \chi(N), g \in F(N)$ and hence replacing X by gX in the above equation, we get

$$\begin{aligned}
\{f_*^{-1}(gX)\}_p h &= (gX)_{f(p)}(h \circ f^{-1}) \\
&= g(f(p))X_{f(p)}(h \circ f^{-1}), \text{ by } (2.26) \\
&= (g \circ f)(p)(f_*^{-1}X)_p h, \text{ from above} \\
\text{or } f_*^{-1}(gX) &= (g \circ f)f_*^{-1}X, \ \forall h.
\end{aligned}$$

Problem 2.53 If f is a transformation of M and g is a differentiable function on M, show that $f_*(gX) = (g \circ f^{-1})f_*X$.

Solution: 53 By virtue of (2.34),

$$\begin{aligned}
\{f_*(gX)_p\}h &= (gX)_p(h \circ f), \ \forall h \in F(M) \\
\text{or } \{f_*(gX)\}_{f(p)} h &= g(p)X_p(h \circ f), \text{ by } (2.26), (2.35) \\
&= g\{(f^{-1}f)(p)\}X_p(h \circ f), \text{ as } f \text{ is a transformation on } M \\
&= \{(g \circ f^{-1})f(p)\}\{f_*(X_p)\}h, \text{ by } (2.34) \\
&= (g \circ f^{-1})f(p)(f_*X)_{f(p)}h
\end{aligned}$$

Thus $f_*(gX) = (g \circ f^{-1})f X, \ \forall h$.

Problem 2.54 If f is a transformation of M and g is a differentiable function on M, prove that $f^*((f_*X)g) = X(f^*g)$.

Solution: 54 Note that $f : M \to M$ is a transformation and hence

$$f(p) = q \Rightarrow p = f^{-1}(q), \quad \forall\, p, q \in M.$$

In view of (2.34), one gets

$$\{f_*(X_p)\}g = X_p(g \circ f), \quad \forall\, g \in F(M)$$
$$\text{or } \{(f_*X)_{f(p)}\}g = \{X(g \circ f)\}(p), \quad \text{by (2.35)}$$
$$\text{or } \{(f_*X)g\}f(p) = \{X(g \circ f)\}(p), \quad \text{by (2.23)}$$
$$\text{or } \{(f_*X)g\}q = \{X(g \circ f)\}f^{-1}(q)$$
$$\text{or } (f_*X)g = \{X(g \circ f)\}f^{-1}\}, \quad \forall\, q$$
$$\text{or } \{(f_*X)g\}f = X(g \circ f)$$
$$\text{or } f^*((f_*X)g) = X(f^*g), \quad \text{by (2.32).}$$

Exercises

Exercise 2.29 *If f is a smooth map from a manifold M into another manifold N and g is a smooth map from N into another manifold L, then prove that $(g \circ f)^* = f^* \circ g^*$.*

Exercise 2.30 *If f is a transformation of M and g is a differentiable function on M, then $f_*[X, Y] = [f_*X, f_*Y]$.*

Geometrical Interpretation of Differential Map

For $X_p \in T_p(M)$, we choose a curve $\sigma(t)$ in M such that X_p is the tangent vector to the curve $\sigma(t)$ at $\sigma(t_0) = p$, $a \le t_0 \le b$ (Fig. 2.20).

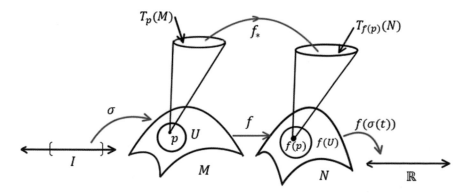

Fig. 2.20 Geometrical interpretation of $f_*(X_p)$

Then $f_*(X_p)$ is defined to be the tangent vector to the curve $f(\sigma(t))$ at $f(p) = f(\sigma(t_0))$ and from (2.15), we have

$$\{f_*(X_p)\}g = \frac{d}{dt}g(f(\sigma(t)))\Big|_{t=t_0}, \quad \forall g \in F(f(p))$$

$$= \frac{d}{dt}(g \circ f)(\sigma(t))\Big|_{t=t_0}$$

$$= X_p(g \circ f), \quad \text{by (2.15)}.$$

Theorem 2.6 *If f is a mapping from an n-dimensional manifold M to an m-dimensional manifold N where $(x^1, x^2, \ldots, x^n)$ is the local coordinate system in a neighbourhood of a point p of M and $(y^1, y^2, \ldots, y^n)$ is the local coordinate system in a neighbourhood of a point $f(p)$ of N, then*

$$f_*\Big(\frac{\partial}{\partial x^i}\Big)_p = \sum_{j=1}^{m}\Big(\frac{\partial f^j}{\partial x^i}\Big)_p\Big(\frac{\partial}{\partial y^j}\Big)_{f(p)}, \quad \text{where } f^j = y^j \circ f.$$

Proof It is known that $\Big\{\frac{\partial}{\partial x^i} : i = 1, 2, 3, \ldots, n\Big\}$ is a basis of $T_p(M)$ and in the same manner $\Big\{\frac{\partial}{\partial y^j} : j = 1, 2, 3, \ldots, m\Big\}$ is a basis of $T_{f(p)}(N)$. Thus

$$f_*\Big(\frac{\partial}{\partial x^i}\Big)_p = \sum_{j=1}^{m} a^j_i\Big(\frac{\partial}{\partial y^j}\Big), \quad i = 1, 2, 3, \ldots, n, \qquad (2.36)$$

where a^j_i's are to be determined. Therefore,

$$\{f_*\Big(\frac{\partial}{\partial x^i}\Big)_p\}y^k = \sum_j a^j_i \delta^k_j = a^k_i.$$

By virtue of (2.34), we obtain

$$\Big(\frac{\partial}{\partial x^i}\Big)_p(y^k \circ f) = a^k_i$$

$$\text{or } \Big(\frac{\partial}{\partial x^i}\Big)_p f^k = a^k_i, \quad \text{by hypothesis}$$

$$\text{or } \Big(\frac{\partial f^k}{\partial x^i}\Big)_p = a^k_i.$$

Using in (2.36), the result follows immediately.

Corollary 2.1 *Let (U, ϕ) be a chart about a point p in a manifold M. If $(u^1, u^2, \ldots, u^n)$ are the standard coordinates of $\mathbb{R}^n$, then*

$$\phi_* \left(\frac{\partial}{\partial x^i}\right)_p = \left(\frac{\partial}{\partial u^i}\right)_{\phi(p)},$$

where $x^i = u^i \circ \phi$, $i = 1, 2, 3, \ldots, n$ are the coordinates of p.

Proof Left to the reader.

Problem 2.55 Let $f : M \to N$ be a diffeomorphism between two smooth manifolds M and N. Then $f_* : T_p(M) \to T_{f(p)}(N)$ is an isomorphism.

Solution: 55 Note that for $f_* : T_p(M) \to T_{f(p)}(N)$, both $T_p(M)$ and $T_{f(p)}(N)$ are vector spaces over $\mathbb{R}$. Hence we have to show

(i) f_* is a linear mapping and
(ii) f_*^{-1} exists.

Now let $f : M \to N$ be a diffeomorphism and hence by definition f^{-1} exists and is of class C^∞. Now $f \circ f^{-1} = I_M$ and $f^{-1} \circ f = I_N$. Again, in view of Problem (2.51), we have

$$(I_M)_* = (f \circ f^{-1})_* = f_* \circ f_*^{-1},$$

where $(I_M)_*$ is the identity differential on $T_p(M)$. Furthermore,

$$(I_N)_* = (f^{-1} \circ f)_* = f_*^{-1} \circ f_*,$$

where $(I_N)_*$ is the identity differential on $T_{f(p)}(N)$. Thus, f_*^{-1} exists and in addition to that it is of class C^∞. Thus f_* is an isomorphism.

Remark 2.28 The matrix representation of f_*, denoted by (f_*), is given by

$$(f_*) = \begin{pmatrix} \frac{\partial f^1}{\partial x^1} & \frac{\partial f^1}{\partial x^2} & \cdots & \frac{\partial f^1}{\partial x^n} \\ \frac{\partial f^2}{\partial x^1} & \frac{\partial f^2}{\partial x^2} & \cdots & \frac{\partial f^2}{\partial x^n} \\ \vdots & \vdots & \vdots & \vdots \\ \frac{\partial f^m}{\partial x^1} & \frac{\partial f^m}{\partial x^2} & \cdots & \frac{\partial f^m}{\partial x^n} \end{pmatrix}. \tag{2.37}$$

Problem 2.56 Find (f_*) where $f : \mathbb{R}^2 \to \mathbb{R}^2$ is given by $f = ((x^1)^2 + (2x^2)^2, 3x^1x^2)$.

Solution: 56 Here $(f_*) = \begin{pmatrix} 2x^1 & 8x^2 \\ 3x^2 & 3x^1 \end{pmatrix}$, where $f^1 = (x^1)^2 + (2x^2)^2$, $f^2 = (3x^1x^2)$.

Exercise

Exercise 2.31 Find (f_*), where

(i) $f : \mathbb{R}^2 \to \mathbb{R}^2$ is defined by $f(x, y) = (xe^y + y, xe^y - y)$.
(ii) $f : \mathbb{R} \to \mathbb{R}$ is defined by $f(x) = e^x$.
(iii) $f : \mathbb{R}^2 \to \mathbb{R}^3$ is defined by $f(x, y) = (x^2y + y^2, x - 2y^3, ye^x)$.
(iv) $f : \mathbb{R}^3 \to \mathbb{R}^2$ is defined by $f(x, y, z) = (x^2 + y^2 + z^2 - 1, ax + by + cz)$.

(v) $f : \mathbb{R}^4 \to \mathbb{R}^2$ is defined by $f(x, y, z, t) = (x^2 + y^2 + z^2 + t^2 - 1, x^2 + y^2 + z^2 + t^2 - 2y - 2z + 5)$.

(vi) $f : \mathbb{R}^2 \to \mathbb{R}^2$ is defined by $f(x, y) = (x^2 + y^2, x^3 y^3)$.

Answers

$$(2.31)(i)\begin{pmatrix} e^y & xe^y + 1 \\ e^y & xe^y - 1 \end{pmatrix} \quad (ii)\ (e^x) \quad (iii)\begin{pmatrix} 2xy & x^2 + 2y \\ 1 & -6y^2 \\ ye^x & e^x \end{pmatrix} \quad (iv)\begin{pmatrix} 2x & 2y & 2z \\ a & b & c \end{pmatrix}$$

$$(v)\begin{pmatrix} 2x & 2y & 2z & 2t \\ 2x & 2y-2 & 2z-2 & 2t \end{pmatrix} \quad (vi)\begin{pmatrix} 2x & 2y \\ 3x^2 y^3 & 3x^3 y^2 \end{pmatrix}.$$

Problem 2.57 If $f : \mathbb{R}^2 \to \mathbb{R}^2$ is given by $(y^1, y^2) = f(x^1, x^2) = ((x^1)^2 + (2x^2)^2, 3x^1 x^2)$, find $f_* \left(\dfrac{\partial}{\partial x^1} \right), \ f_* \left(\dfrac{\partial}{\partial x^2} \right)$.

Solution: 57 From Theorem 2.6, we see that

$$f_* \left(\frac{\partial}{\partial x^i} \right) = \sum_{j=1}^{2} \left(\frac{\partial f^j}{\partial x^i} \right) \frac{\partial}{\partial y^j}, \quad i = 1, 2.$$

$$\therefore \ f_* \left(\frac{\partial}{\partial x^1} \right) = \left(\frac{\partial f^1}{\partial x^1} \right) \frac{\partial}{\partial y^1} + \left(\frac{\partial f^2}{\partial x^1} \right) \frac{\partial}{\partial y^2}$$

$$= 2x^1 \frac{\partial}{\partial y^1} + 3x^2 \frac{\partial}{\partial y^2} \quad \text{and}$$

$$f_* \left(\frac{\partial}{\partial x^2} \right) = \left(\frac{\partial f^1}{\partial x^2} \right) \frac{\partial}{\partial y^1} + \left(\frac{\partial f^2}{\partial x^2} \right) \frac{\partial}{\partial y^2}$$

$$= 4x^2 \frac{\partial}{\partial y^1} + 3x^1 \frac{\partial}{\partial y^2}.$$

Alternative

Here

$$f_* \left(\frac{\partial}{\partial x^1} \right) = \begin{pmatrix} \frac{\partial f^1}{\partial x^1} & \frac{\partial f^1}{\partial x^2} \\ \frac{\partial f^2}{\partial x^1} & \frac{\partial f^2}{\partial x^2} \end{pmatrix} \begin{pmatrix} 1 \\ 0 \end{pmatrix} = \begin{pmatrix} 2x^1 & 4x^2 \\ 3x^2 & 3x^1 \end{pmatrix} \begin{pmatrix} 1 \\ 0 \end{pmatrix} = \begin{pmatrix} 2x^1 \\ 3x^2 \end{pmatrix}.$$

Since the vector $f_* \left(\dfrac{\partial}{\partial x^1} \right)$ is the linear combination of the basis vectors $\left\{ \dfrac{\partial}{\partial y^1}, \dfrac{\partial}{\partial y^2} \right\}$, we write from above

$$f_* \left(\frac{\partial}{\partial x^1} \right) = 2x^1 \frac{\partial}{\partial y^1} + 3x^2 \frac{\partial}{\partial y^2}$$

$$\text{Similarly,} \quad f_* \left(\frac{\partial}{\partial x^2} \right) = \begin{pmatrix} 1 \\ 0 \end{pmatrix} = \begin{pmatrix} 2x^1 & 4x^2 \\ 3x^2 & 3x^1 \end{pmatrix} \begin{pmatrix} 0 \\ 1 \end{pmatrix} = \begin{pmatrix} 4x^2 \\ 3x^1 \end{pmatrix}$$

$$\therefore \ f_* \left(\frac{\partial}{\partial x^2} \right) = 4x^2 \frac{\partial}{\partial y^1} + 3x^1 \frac{\partial}{\partial y^2}.$$

Exercises

Exercise 2.32 *A. Find (f_*) at $(0, 0)$ in Exercise 2.31(i), (ii) and (iii).*
B. Find (f_*) at $(0, 0, 0)$ in Exercise 2.31(iv).
C. Find (f_*) at $(0, 0, 0, 0)$ in Exercise 2.31(v).

Exercise 2.33 *A. Let $f : \mathbb{R}^2 \to \mathbb{R}^2$ be defined by*
(i) $f(x, y) = (x^2 - 2y, 4x^3 y^2)$. Find $(f_*)_{(1,2)}$.
(ii) $f(x, y) = (x^2 + y^2, x^3 y^3)$. Find $(f_*)_{(2,1)}$.
B. Let $g : \mathbb{R}^2 \to \mathbb{R}^3$ be defined by $g(x, y) = (x^2 y - y, 2x^3 - y, xe^y)$. Find $(g_*)_{(2,1)}$.

Exercise 2.34 *A. Find $(f_*)\left(\dfrac{\partial}{\partial x}\right), (f_*)\left(\dfrac{\partial}{\partial y}\right), (f_*)\left(\dfrac{\partial}{\partial z}\right), (f_*)\left(\dfrac{\partial}{\partial t}\right)$ where*
(i) $f : \mathbb{R}^2 \to \mathbb{R}^2$ is defined by $(u, v) = f(x, y) = (xe^y + y, xe^y - y)$.
(ii) $f : \mathbb{R}^2 \to \mathbb{R}^3$ is defined by $(u, v, w) = f(x, y) = (x^2 y + y^2, x - 2y^3, ye^x)$.
(iii) $f : \mathbb{R}^3 \to \mathbb{R}^2$ is defined by $(u, v) = f(x, y, z) = (x^2 + y^2 + z^2 - 1, ax + by + cz)$.
(iv) $f : \mathbb{R}^4 \to \mathbb{R}^2$ is defined by $(u, v) = f(x, y, z, t) = (x^2 + y^2 + z^2 + t^2 - 1, x^2 + y^2 + z^2 + t^2 - 2y - 2z + 5)$.
B. Find $(f_*)\left(\dfrac{\partial}{\partial x}\right)$ where $f : \mathbb{R}^2 \to \mathbb{R}^3$ is defined by $(u, v, w) = f(x, y)$ $= (x, y, xy)$.

Answers

2.32 A. (i) $\begin{pmatrix} 1 & 1 \\ 1 & -1 \end{pmatrix}$ (ii) (1) (iii) $\begin{pmatrix} 0 & 0 \\ 1 & 0 \\ 0 & 1 \end{pmatrix}$ B. $\begin{pmatrix} 0 & 0 & 0 \\ a & b & c \end{pmatrix}$ C. $\begin{pmatrix} 0 & 0 & 0 & 0 \\ 0 & -2 & -2 & 0 \end{pmatrix}$.

2.33 A. (i) $\begin{pmatrix} 2 & -2 \\ 48 & 16 \end{pmatrix}$ (ii) $\begin{pmatrix} 4 & 1 \\ 12 & 24 \end{pmatrix}$ B. $\begin{pmatrix} 4 & 0 \\ 24 & -1 \\ e & 2e \end{pmatrix}$.

2.34 A. (i) $e^y \frac{\partial}{\partial u} + e^y \frac{\partial}{\partial v}, \quad (xe^y + 1)\frac{\partial}{\partial u} + (xe^y - 1)\frac{\partial}{\partial v}$.
(ii) $2xy \frac{\partial}{\partial u} + \frac{\partial}{\partial v} + ye^x \frac{\partial}{\partial w}, \quad (x^2 + 2y)\frac{\partial}{\partial u} - 6y^2 \frac{\partial}{\partial v} + e^x \frac{\partial}{\partial w}$.
(iii) $2x \frac{\partial}{\partial u} + a \frac{\partial}{\partial v}, 2y \frac{\partial}{\partial u} + b \frac{\partial}{\partial v}, \quad 2z \frac{\partial}{\partial u} + c \frac{\partial}{\partial v}$.
(iv) $2x \frac{\partial}{\partial u} + 2x \frac{\partial}{\partial v}, \quad 2y \frac{\partial}{\partial u} + 2(y - 1)\frac{\partial}{\partial v}, \quad 2z \frac{\partial}{\partial u} + 2(z - 1)\frac{\partial}{\partial v}, \quad 2t \frac{\partial}{\partial u} + 2t \frac{\partial}{\partial v}$.
B. $\frac{\partial}{\partial u} + y \frac{\partial}{\partial w}, \quad \frac{\partial}{\partial v} + x \frac{\partial}{\partial w}$.

Problem 2.58 Let $f : \mathbb{R}^3 \to \mathbb{R}$ be defined by $f(x, y, z) = x^2 y$. If $X = xy \dfrac{\partial}{\partial x} + x^2 \dfrac{\partial}{\partial z}$; compute $f_*(X)_{(1,1,0)}$ OR $(f_*)_{(1,1,0)} (X)_{(1,1,0)}$.

Solution: 58 Taking into consideration (2.35), we know that

$$f_*(X)_{(1,1,0)} = (f_* X)_{f(1,1,0)} = (f_* X)_1.$$

Now

$$f_*(X)_{(1,1,0)} = \left(\frac{\partial f}{\partial x} \; \frac{\partial f}{\partial y} \; \frac{\partial f}{\partial z}\right)_{(1,1,0)} \begin{pmatrix} xy \\ 0 \\ x^2 \end{pmatrix}_{(1,1,0)}$$

$$= (2 \; 1 \; 0) \begin{pmatrix} 1 \\ 0 \\ 1 \end{pmatrix}$$

$$= 2.$$

Thus, $f_*(X)_{(1,1,0)} = 2\frac{d}{dt}\big|_1$, where t denotes the canonical coordinate of $\mathbb{R}$.

Exercises

Exercise 2.35 If $X = x^2\frac{\partial}{\partial y}$, compute $(f_*)_{(1,1)}(X)_{(1,1)}$ for Exercise 2.31 (i), (iii).

Exercise 2.36 Let $f : \mathbb{R}^2 \to \mathbb{R}^2$ be defined by $f(x, y) = (x^2 + y, x^3 y^3)$ and $g : \mathbb{R}^2 \to \mathbb{R}^3$ be defined by $g(x, y) = (x^2 y - y, 2x^3 - y, xe^y)$. Compute the following:
(i) $f_*\left(\left(2\frac{\partial}{\partial x} + \frac{\partial}{\partial y}\right)_{(0,1)}\right)$ (ii) $g_*\left(\left(2\frac{\partial}{\partial x} + \frac{\partial}{\partial y}\right)_{(0,1)}\right)$.

Answers
2.35. (i) $(e + 1)\frac{\partial}{\partial u}\big|_{(e+1,e-1)} + (e - 1)\frac{\partial}{\partial v}\big|_{(e+1,e-1)}$ (ii) $(3\frac{\partial}{\partial u} - 6\frac{\partial}{\partial v} + e\frac{\partial}{\partial w})_{(2,-1,e)}$.
2.36. (i) $(\frac{\partial}{\partial x})_{(1,0)}$ (ii) $(-\frac{\partial}{\partial x} - \frac{\partial}{\partial y} + 2e\frac{\partial}{\partial z})_{(-1,-1,0)}$.

Problem 2.59 Let $f : \mathbb{R}^2 \to \mathbb{R}^2$ be defined by

$$(u, v) = f(x_1, x_2) = \begin{pmatrix} \cos\theta & -\sin\theta \\ \sin\theta & \cos\theta \end{pmatrix} \begin{pmatrix} x_1 \\ x_2 \end{pmatrix}.$$

Let $X = -x_2\frac{\partial}{\partial x_1} + x_1\frac{\partial}{\partial x_2}$ be a vector field on $\mathbb{R}^2$. If $p = (x_1, x_2) \in \mathbb{R}^2$ and

$$f_*X_p = \left(a\frac{\partial}{\partial u} + b\frac{\partial}{\partial v}\right)_{f(p)},$$

find a, b.

Solution: 59 Here $(u, v) = f(x_1, x_2) = (x_1 \cos\theta - x_2 \sin\theta, \; x_1 \sin\theta + x_2 \cos\theta)$. Now

$$\{f_* X_p\}u = \left\{a \frac{\partial}{\partial u} + b \frac{\partial}{\partial v}\right\} = a$$

or, $X_p(u \circ f) = a$

or, $\left(-x_2 \dfrac{\partial}{\partial x_1} + x_1 \dfrac{\partial}{\partial x_2}\right)(x_1 \cos\theta - x_2 \sin\theta) = a$

or, $-x_2 \cos\theta - x_1 \sin\theta = a$.

Similarly, $\left(-x_2 \dfrac{\partial}{\partial x_1} + x_1 \dfrac{\partial}{\partial x_2}\right)(x_1 \sin\theta + x_2 \cos\theta) = b$

or, $-x_2 \sin\theta + x_1 \cos\theta = b$.

The values of a, b are therefore calculated.

Alternative

Here $(f_*) = \begin{pmatrix} \cos\theta & -\sin\theta \\ \sin\theta & \cos\theta \end{pmatrix}_{p=(x_1, y)}$ and $(X)_p = \begin{pmatrix} -x_2 \\ x_1 \end{pmatrix}_{(x_1, x_2)}$. So

$$(f_*)(X)_p = \begin{pmatrix} \cos\theta & -\sin\theta \\ \sin\theta & \cos\theta \end{pmatrix}\begin{pmatrix} -x_2 \\ x_1 \end{pmatrix}_{f(p)=f(x_1,x_2)}$$

$$= \begin{pmatrix} -x_2 \cos\theta - x_1 \sin\theta \\ -x_2 \sin\theta + \cos\theta \end{pmatrix}_{(x_1 \cos\theta - x_2 \sin\theta,\, x_1 \sin\theta + x_2 \cos\theta)}$$

$$= \left\{(-x_2 \cos\theta - x_1 \sin\theta)\frac{\partial}{\partial u} + (-x_2 \sin\theta + \cos\theta)\frac{\partial}{\partial u}\right\}_{f(p)}.$$

Note that, for the linear map $f_* : T_p(M) \to T_{f(p)}(N)$ at the point $p \in M$, the Kernel of f_* at p is given by

$$\ker f_* = \{X_p \in T_p(M) | f_*(X_p) = 0,\ 0 \in T_{f(p)}(N)\}.$$

Here $\ker f_*$ is a subspace of $T_p(M)$. Also, the image of f_* at p is

$$\text{Image } f_* = \{Y_{f(p)} \in T_{f(p)}(N) | f_*(X_p) = Y_{f(p)}\},$$

which is a subspace of $T_{f(p)}(N)$.

Problem 2.60 Let $f : \mathbb{R}^4 \to \mathbb{R}^2$ be defined by $f(x_1, x_2, x_3, x_4) = (u, v) = (x_1^2 + x_2^2 + x_3^2 + x_4^2 - 1,\ x_1^2 + x_2^2 + x_3^2 + x_4^2 - 2x_2 - 2x_3 + 5)$.

(i) Find a basis of $\ker f_*$ at $(0, 1, 2, 0)$.
(ii) Find the image by (f_*) of $(1, 0, 2, 1) \in T_{(1,2,0,1)}\mathbb{R}^4$.

Solution: 60 (i) As $f : \mathbb{R}^4 \to \mathbb{R}^2$ is defined by $f(x_1, x_2, x_3, x_4) = (x_1^2 + x_2^2 + x_3^2 + x_4^2 - 1,\ x_1^2 + x_2^2 + x_3^2 + x_4^2 - 2x_2 - 2x_3 + 5)$, then for $p = (0, 1, 2, 0)$, $f_* : T_{(0,1,2,0)}(\mathbb{R}^4) \to T_{(4,4)}(\mathbb{R}^2)$ is a differential map such that $\ker f_*$ at $(0, 1, 2, 0)$ is a vector subspace of $T_{(0,1,2,0)}(\mathbb{R}^4)$.

Now any $X \in \chi(\mathbb{R}^4)$ can be expressed as $X = a\dfrac{\partial}{\partial x_1} + b\dfrac{\partial}{\partial x_2} + c\dfrac{\partial}{\partial x_3} + d\dfrac{\partial}{\partial x_4}$, where $a, b, c, d \in \mathbb{R}$. Here

$$(f_*) = \begin{pmatrix} 2x_1 & 2x_2 & 2x_3 & 2x_4 \\ 2x_1 & 2(x_2 - 1) & 2(x_3 - 1) & 2x_4 \end{pmatrix}.$$

Therefore $(f_*)_{(0,1,2,0)} = \begin{pmatrix} 0 & 2 & 4 & 0 \\ 0 & 0 & 2 & 0 \end{pmatrix}$. Thus, $(f_*)_{(0,1,2,0)} X \in T_{f(p)}(\mathbb{R}^2)$ is a linear combination of $\{\dfrac{\partial}{\partial u}, \dfrac{\partial}{\partial v}\}$, where $p = (0, 1, 2, 0)$, i.e. $(f_*)_{(0,1,2,0)} X_p = (2b + 4c)\dfrac{\partial}{\partial u} + 2c\dfrac{\partial}{\partial v}$. But $\ker f_*$ is such that $(f_*)_{(0,1,2,0)} X_p = \theta, \theta \in T_{(4,4)}(\mathbb{R}^2)$, where $p = (0, 1, 2, 0)$. Consequently, we must have $b = 0 = c$. Thus

$$X_p = a\dfrac{\partial}{\partial x_1} + d\dfrac{\partial}{\partial x_2}, \quad p = (0, 1, 2, 0).$$

Consequently,

$$\ker(f_*)_{(0,1,2,0)} = \{X_{(0,1,2,0)} \in T_{(0,1,2,0)}(\mathbb{R}^4) \,|\, (f_*)_{(0,1,2,0)} X_{(0,1,2,0)} = \theta\},$$

and the basis is $\{(\dfrac{\partial}{\partial x_1})_{(0,1,2,0)}, (\dfrac{\partial}{\partial x_2})_{(0,1,2,0)}\}$ such that $\ker(f_*)_{(0,1,2,0)}$ is a subspace of $T_{(0,1,2,0)}(\mathbb{R}^4)$.

(ii) Also $(f_*)_{(1,2,0,1)} = \begin{pmatrix} 2 & 4 & 0 & 2 \\ 2 & 2 & -2 & 2 \end{pmatrix}$ and $f(1, 2, 0, 1) = (5, 7)$. Thus, $(f_*)_{(1,2,0,1)} X_p \in T_{(5,7)}(\mathbb{R}^2)$ can be expressed as

$$(f_*)_{(1,2,0,1)} X_p = (2a + 4b + 2d)\dfrac{\partial}{\partial u}\Big|_{f(p)} + (2a + 2b - 2c + 2d)\dfrac{\partial}{\partial v}\Big|_{f(p)}, \quad p = (1, 2, 0, 1) \text{ and } X = (1, 0, 2, 1).$$

Therefore $(f_*)_{(1,2,0,1)} X_p = 4\dfrac{\partial}{\partial u}\Big|_{(5,7)}$.

Problem 2.61 Let $f : \mathbb{R}^2 \to \mathbb{R}^3$ be defined by $f(x, y) = (x^2 y - y, 2x^3 - y, xe^y)$. Calculate the conditions that the constants A, B, C must satisfy for the vector

$$\left(A\dfrac{\partial}{\partial x} + B\dfrac{\partial}{\partial y} + C\dfrac{\partial}{\partial z} \right)_{f(0,0)}$$

to be the image of some vector by f_*.

Solution: 61 Note that $(f_*)_{(0,0)} = \begin{pmatrix} 0 & -1 \\ 0 & -1 \\ 1 & 0 \end{pmatrix}$. Let us choose $X = \lambda\dfrac{\partial}{\partial x} + \mu\dfrac{\partial}{\partial y} \in$

$\chi(\mathbb{R}^2)$ such that

$$(f_*)_{(0,0)} X_{(0,0)} = \begin{pmatrix} 0 & -1 \\ 0 & -1 \\ 1 & 0 \end{pmatrix} \begin{pmatrix} \lambda \\ \mu \end{pmatrix} = \begin{pmatrix} -\mu \\ -\mu \\ \lambda \end{pmatrix}_{f(0,0)} = \left(A\dfrac{\partial}{\partial x} + B\dfrac{\partial}{\partial y} + C\dfrac{\partial}{\partial z}\right)_{f(0,0)}.$$

Compare $A = -\mu$, $B = -\mu$, $C = \lambda$. Note that $(f_*)_{(0,0)} : T_{(0,0)}(\mathbb{R}^2) \to T_{(0,0,0)}(\mathbb{R}^3)$. Here Image f_* is a subspace of $T_{(0,0,0)}(\mathbb{R}^3)$ by definition. Thus, the image of f_* of $T_{(0,0)}(\mathbb{R}^2)$ is a vector subspace of $T_{(0,0,0)}(\mathbb{R}^3)$ of vectors of type $(-A, -A, C)$.

Exercises

Exercise 2.37 A. *Let* $f : \mathbb{R}^4 \to \mathbb{R}^2$ *be defined by*

$$f(x_1, x_2, x_3, x_4) = (u, v) = (x_1^2 + x_2^2 + x_3^2 + x_4^2 + 1, x_1^2 + x_2^2 + x_3^2 + x_4^2 - 2x_1 - 2x_4 + 6).$$

(i) *Find the basis of* $\ker f_*$ *at* $(1, 0, 1, 0)$.
(ii) *Find the image by* (f_*) *of* $(1, 0, 1, 1) \in T_{(1,1,0,1)}\mathbb{R}^2$.
B. *Let* $g : \mathbb{R}^2 \to \mathbb{R}^2$ *be defined by* $g(x_1, x_2) = (x_1^2 + x_2, x_1^3 x_2^3)$. *Calculate the conditions that the constants* A, B *must satisfy for the vector*

$$\left(A\dfrac{\partial}{\partial x_1} + B\dfrac{\partial}{\partial x_2}\right)_{g(0,0)}$$

to be the image of some vector by g_*.

Exercise 2.38 *Let us fix* θ *and define* $f : \mathbb{R}^4 \to \mathbb{R}^4$ *by* $f_\theta(x, y, z, t) = \begin{pmatrix} \cos\theta & -\sin\theta \\ \sin\theta & \cos\theta \end{pmatrix} \begin{pmatrix} x & z \\ y & t \end{pmatrix}$.

(i) *Compute* $(f_\theta)_*$.
(ii) *Compute* $(f_\theta)_* X$, *where* $X = \cos\theta\dfrac{\partial}{\partial x} - \sin\theta\dfrac{\partial}{\partial y} + \cos\theta\dfrac{\partial}{\partial z} - \sin\theta\dfrac{\partial}{\partial t}$.

Answers

2.37 A. (i) $\{(\frac{\partial}{\partial y})_{(1,0,1,0)}, (\frac{\partial}{\partial w})_{(1,0,1,0)}\}$ (ii) $4\frac{\partial}{\partial u}\big|_{(4,7)}$ B. $(A, 0)$.

2.38 (i) $\begin{pmatrix} \cos\theta & -\sin\theta & 0 & 0 \\ \sin\theta & \cos\theta & 0 & 0 \\ 0 & 0 & \cos\theta & \sin\theta \\ 0 & 0 & \sin\theta & \cos\theta \end{pmatrix}$ (ii) $\frac{\partial}{\partial x} + \frac{\partial}{\partial z}$.

A smooth map $f : M \to N$ is said to be a smooth submersion (or simply submersion) at $p \in M$, if f_* at p, i.e. $f_{*,p}$ is surjective. Equivalently, we can say that

$$\dim \text{Image } f_{*,p} = \dim T_{f(p)}(N) = m.$$

Moreover, f is called a smooth immersion (or simply immersion) at $p \in M$, if its differential $f_{*,p}$ is injective. In other words, we have

$$\ker f_{*,p} = \{\theta\} \implies \dim \ker f_{*,p} = 0.$$

Hence, using the well-known theorem in linear algebra, viz.

$$\dim \ker f_{*,p} + \dim \text{Image } f_{*,p} = \dim M = n,$$

we find $\dim \text{Image } f_{*,p} = n$. Thus, if a smooth map $f : M^n \to N^m$ is a submersion at a point p in M, then $n \geq m$ and if immersion at a point p in M then $n \leq m$. A simple question arises: Is an injective or surjective map f, respectively, an immersion or submersion? For this, we need to explain immersion and submersion using the rank of the differential map f_* at the point p in M.

We know that the dimension of Image $f_{*,p}$ is said to be the **rank** of f_* at p, denoted by rank $f_{*,p}$. We define the rank of f at p, denoted by rank $f_{|p}$, to be of r if rank f_* at $p = r$. In other words, we can also say that the rank f_* is the rank of the Jacobian matrix of f with respect to any smooth chart. If f has the same rank r at every point of M, we say that it has constant rank r, and write rank $f = r$.

Since rank $f_{*,p} \leq \min\{\dim T_p(M), \dim T_{f(p)}(N)\} = \min\{n, m\}$, therefore

$$\text{rank } f_{|p} \leq \min\{n, m\}.$$

A smooth map $f : M \to N$ is said to be a submersion at $p \in M$, if

$$\text{rank } f_{*,p} = \dim \text{Image } f_{*,p} = \dim T_{f(p)}(N) = m \ i.e. \ \text{rank } f_{|p} = m.$$

Moreover, f is called an immersion at $p \in M$, if

$$\text{rank } f_{|p} = \dim M = n.$$

If we consider a smooth map $f : \mathbb{R}^2 \to \mathbb{R}^3$, then its differential at $p \in \mathbb{R}^2$ is given by

$$f_{*,p} : T_p(\mathbb{R}^2) \to T_{f(p)}(\mathbb{R}^3).$$

Here

$$\text{rank } f_{*,p} \leq \min\{2, 3\} = 2 \Rightarrow \text{rank } f_{*,p} = 1 \text{ or } 2.$$

Here, f fails to be submersion at p, for if f is submersion at p then rank $f_* = 3$ which is not possible. Thus we can say that if

$$\dim M = n < \dim N = m,$$

then f fails to be a submersion at any point p in M. However, if rank $f_* = 2$, then rank $f_{|p} = 2$ implies f is immersion at p in $\mathbb{R}^2$. In a similar manner for a smooth map

$f : \mathbb{R}^3 \to \mathbb{R}^2$, to be immersion at $p \in \mathbb{R}^3$, we have rank $f_* = 3$, i.e. rank $f_{|p} = 3$ which is not possible. Thus if

$$\dim M = n > \dim N = m,$$

then f fails to be an immersion at any point p in M.

A smooth map $f : M \to N$ is said to be a submersion and immersion if its differential is respectively surjective and injective at every point of M.

Remark 2.29 Let $f : M \to N$ be a smooth map of constant rank. Then
(i) if f is surjective, it is a submersion.
(ii) if f is injective, it is an immersion.
(iii) if f is bijective, it is a diffeomorphism.

Examples

Example 2.9 Let us consider the curve $\gamma : (-\pi, \pi) \to \mathbb{R}^2$ defined by $\gamma(t) = (\sin 2t, \sin t)$. Here, γ is injective but $\gamma'(t)$ does not vanish for any t. Hence γ is an injective immersion.

Example 2.10 Let us consider the function $f : \mathbb{R} \to \mathbb{R}^2$ by $t \mapsto (t^2 - 1, t(t^2 - 1))$. Here f is not injective as $f(1) = f(-1) = (0, 0)$. But f' does not vanish for any t, so f is an immersion but not injective.

Example 2.11 Suppose $M_1, M_2, \ldots, M_s$ are the smooth manifolds. Then each of the projection maps

$$\pi_i : M_1 \times M_2 \times \cdots \times M_s \to M_i$$

is a submersion. In particular, if π is a projection map from $\mathbb{R}^{n+k}$ to $\mathbb{R}^n$ by

$$(x^1, x^2, \ldots, x^m, x^{m+1}, \ldots, x^n) \mapsto (x^1, x^2, \ldots, x^m),$$

then π is a submersion.

Example 2.12 If U is an open subset of a manifold M, then the inclusion map $i : U \hookrightarrow M$ is both an immersion and submersion. Moreover, here the map is not surjective. So this example shows that a submersion need not be surjective.

Example 2.13 Let us consider the map $f : \mathbb{R} \to \mathbb{R}$, $f(x) = x^3$. Here f is surjective but at $x = 0$, $df = f'(x) = 3x^2$ is not surjective. Hence f fails to be a submersion. So this example shows that a surjective map need not be a submersion.

Let $f : M \to N$ be a smooth map. A point $p \in M$ is said to be a **critical point** of f if $f_{*,p}$ is not surjective. A point $q \in N$ is said to be a **critical value** of f if the set $f^{-1}(q)$ contains a critical point of f. In other words, a point in N is a **critical value** if it is the image of some critical point in M.

In particular, let $f : M \to \mathbb{R}$ be a smooth map on M. A point $p \in M$ is said to be a **critical point** of f if $f_{*,p} = 0$.

Proposition 2.8 *Let* $f : N^n \to \mathbb{R}$ *be a* C^∞ *map. A point* $p \in N$ *is a* **critical point** *if and only if it is relative to a coordinate system* $(U, x^1, x^2, x^3, \ldots, x^n)$ *at* $p \in N$,
$$\frac{\partial f}{\partial x^i}\Big|_p = 0.$$

Proof For $f : N^n \to \mathbb{R}$, the map $f_* : T_p(N) \to T_{f(p)}(\mathbb{R})$ is such that

$$(f_*)_p = \left(\frac{\partial f}{\partial x^1} \frac{\partial f}{\partial x^2} \cdots \frac{\partial f}{\partial x^n}\right)_p.$$

It is known from the previous discussion that Image f_* is a subspace of $T_{f(p)}(\mathbb{R})$.

Thus it is either 0-dimensional or 1-dimensional. Hence either f_* is a zero-map or a surjective map.

Thus, f_* will not be surjective if and only if $\frac{\partial f}{\partial x^i}\Big|_p = 0$.

It means that the real number $f(p)$ is called a critical value of f. A critical point is called **non-degenerate** if

$$\det\left(\frac{\partial^2 f}{\partial x^i \partial x^j}(p)\right) \neq 0.$$

Remark 2.30 Non-degeneracy is independent of the choice of coordinate system.

A point $p \in M$ is a **regular point** of $f : M \to N$ if f_*, p is **surjective**. In other words, we can say that $p \in M$ is a **regular point** of f if and only if f is a **submersion** at p, i.e. rank $f_* = \dim M$.

Example 2.14 The function $f(x) = x + e^{-x}$ has a critical point at $c = 0$. The derivative is zero at this point. So

$$f'(x) = (x + e^{-x})' = 1 - e^{-x}.$$

Now $f'(c) = 1 - e^{-c} = 0 \Rightarrow c = 0$.

Example 2.15 The function $f(x) = 2x - x^2$ has a critical point at $c = 1$. The derivative is zero at this point. Here $f'(x) = 2 - 2x$. So, $f'(c) = 0 \Rightarrow c = 1$.

Problem 2.62 Find the critical points of the map $f : \mathbb{R}^3 \to \mathbb{R}^2$ given by

$$(x, y, z) \mapsto (xz, y).$$

Solution: 62 Let $p \in \mathbb{R}^3$. The Jacobian matrix $f_{*,p}$ is given by

$$f_{*,p} = \begin{pmatrix} \frac{\partial}{\partial x}(xz) & \frac{\partial}{\partial y}(xz) & \frac{\partial}{\partial z}(xz) \\ \frac{\partial}{\partial x}(y) & \frac{\partial}{\partial x}(y) & \frac{\partial}{\partial x}(y) \end{pmatrix} = \begin{pmatrix} z & 0 & x \\ 0 & 1 & 0 \end{pmatrix}.$$

Here $f_{*,p}$ fails to be surjective if rank $f_* < 2$ if and only if $x = z = 0$. Hence the set of critical points of f is the y-axis.

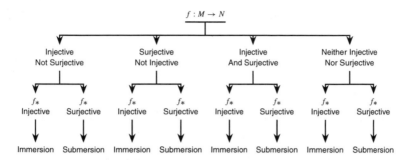

2.13 Submanifolds

Let us consider a map $f : M \to N$, where M and N are topological manifolds. f is said to be a **topological embedding (topological imbedding)** if f is a homeomorphism onto its image $f(M)(\subset N)$, where $f(M)$ is endowed with the subspace topology inherited from N.

The smooth map $f : M \to N$, where M and N are smooth manifolds, is a **smooth embedding (or smooth imbedding)** if f is an immersion together with topological embedding.

Let N be a smooth manifold and $M \subseteq N$. Let M be a manifold endowed with the subspace topology inherited from N. Then M is said to be an **embedded submanifold (or regular submanifold)** of N if M is endowed with a smooth structure with respect to which the inclusion map of M in N is a smooth embedding.

Let N be an m-dimensional smooth manifold and $M \subset N$. Let M be a manifold equipped with a topology, not necessarily the subspace topology inherited from N, with respect to which it is a topological manifold of dimension n. Then M is said to be an **immersed submanifold** of N if M is endowed with a smooth structure with respect to which the topological manifold M becomes an n-dimensional smooth manifold such that the inclusion map $i : M \hookrightarrow N$ is a smooth immersion. Also, the immersed submanifold M has co-dimension $m - n$.

It is evident that every embedded submanifold is an immersed submanifold (refer to Problem 2.63). For the sake of simplicity, embedded submanifold and immersed submanifold are always of the smooth kind.

Example 2.16 If $M_1, M_2, \ldots, M_s$ are the smooth manifolds and $q_i \in M_i$ are arbitrary points, each of the maps

$$\zeta_i : M_i \to M_1 \times M_2 \times \cdots \times M_s,$$

given by $\zeta_i(p) = (q_1, q_2, \ldots, q_{j-1}, p, q_{j+1}, \ldots, q_s)$, is a smooth embedding.

Example 2.17 The map $\gamma : \mathbb{R} \to \mathbb{R}^2$ given by $\gamma(t) = (t^5, 0)$ is a smooth map but is not a topological embedding because $\gamma'(0) = 0$.

Example 2.18 Let $f : \mathbb{R} \to \mathbb{R}^2$ be defined by $f(t) = (\cos \pi t, \sin \pi t)$, $\forall\, t \in \mathbb{R}$. Here f is an immersion but not an injective map. So it fails to be a topological embedding.

Example 2.19 The circle S^1 is a 1-dimensional embedded submanifold of $\mathbb{R}^2$.

Example 2.20 The graph $y = (x)^{\frac{1}{3}}$ is an embedded submanifold of $\mathbb{R}^2$.

Example 2.21 The sphere S^n is an embedded submanifold of $\mathbb{R}^n$ with dimension $n - 1$.

Example 2.22 Let $f : \mathbb{R}^m \to \mathbb{R}^n$ be a smooth map. Then its graph

$$\gamma = \{(x_1, x_2, \ldots, x_m, y_1, y_2, \ldots, y_n) \in \mathbb{R}^{m+n} \,\big|\, f(x_1, x_2, \ldots, x_m) = (y_1, y_2, \ldots, y_n)\}$$

is a smooth m-dimensional embedded submanifold of $\mathbb{R}^{m+n}$.

Lemma 2.1 *Let X, Y be topological spaces. Let $F : X \to Y$ be $1 - 1$, continuous and open map. Then F is a topological embedding.*

Proof Let U be an open subset of X. Since F is open, $F(U)$ is open in Y. Hence $F(U) \cap F(X)$ is open in $F(X)$. Since $F(U) \subset F(X)$, therefore $F(U)$ is open in $F(X)$. This proves F is a topological embedding.

Lemma 2.2 *Let X, Y be topological spaces. Let $F : X \to Y$ be $1 - 1$, continuous and closed map. Then F is a topological embedding.*

Proof Left to the reader.

Lemma 2.3 *Closed map lemma: Let X be a compact space and Y be a Hausdorff space. Let $F : X \to Y$ be a $1 - 1$, continuous map. Then F is a topological embedding.*

Proof Let A be any closed subset of X. Since A is closed in the compact space X, A is compact in X. Since F is continuous, therefore $F(A)$ is compact in Y, and Y being Hausdorff, therefore $F(A)$ is closed in Y. This shows F is a closed map. Since $F : X \to Y$ is a $1 - 1$, continuous map, therefore F is a topological embedding (refer to Lemma 2.2).

Now we are going to define a proper map between two topological spaces as follows.

Let X, Y be topological spaces. Let $F : X \to Y$ be a mapping. If for every compact subset W of Y, the inverse image $F^{-1}(W)$ is compact in X, then we say that $F : X \to Y$ is a proper map.

Lemma 2.4 *Let X be a Hausdorff topological space and Y be a Hausdorff locally compact space. Let $F : X \to Y$ be a continuous map. Let F be a proper mapping. Then F is a closed mapping.*

Proof Left to the reader.

Lemma 2.5 *Let M be an n-dimensional smooth manifold and N be an m-dimensional smooth manifold. Let $F : M \to N$ be a $1-1$ smooth immersion. If F is a proper mapping, then $F : M \to N$ is a smooth embedding.*

Proof Since M is an n-dimensional smooth manifold, M is a Hausdorff locally compact space. Similarly, N is a Hausdorff locally compact space. Since F is a smooth map, it is continuous. Further, since F is a proper map, F is a closed map (refer to Lemma 2.4). Since F is a $1-1$ smooth immersion, and F is closed, therefore F is a smooth embedding.

Lemma 2.6 *Let M be an n-dimensional compact smooth manifold and N be an m-dimensional smooth manifold. If $F : M \to N$ is a $1-1$ smooth immersion, then $F : M \to N$ is proper mapping.*

Proof Let W be any compact subset of N. Our claim is to show that the inverse image $F^{-1}(W)$ is compact in M. Since W is compact in N and N is Hausdorff, therefore W is closed in N. Since F is smooth, it is continuous. Hence $F^{-1}(W)$ is closed in M. As M is compact, $F^{-1}(W)$ is compact in M. This proves F is proper, hence a smooth embedding (refer to Lemma 2.5).

Problem 2.63 Let M be an n-dimensional smooth manifold and $S(\neq \phi) \subset M$. Let S be an embedded submanifold of M with co-dimension k, k $= 1, 2, 3, ..., $ n $- 1$. Prove that S is an immersed submanifold of M with co-dimension k.

Solution: 63 Let us set $\tau_s = \{G \cap S : G$ is open in $M\}$. Since S is an embedded submanifold of M with co-dimension k, τ_s is a topology over S (called the subspace topology of S) with respect to which S becomes a $(n - k)$-dimensional topological manifold. Moreover, $\exists$ an C^∞-atlas $\mathcal{A}$ on S, with respect to which the topological manifold S becomes an $(n - k)$-dimensional smooth manifold such that the map $i : S \hookrightarrow M$ is a smooth embedding, and hence i is a smooth immersion. Thus, S is an immersed submanifold of M with co-dimension $n - (n - k) = k$. This completes the solution.

Problem 2.64 Let M be an n-dimensional smooth manifold with C^∞-atlas $\mathcal{A}$ and N be an m-dimensional smooth manifold with C^∞-atlas $\mathcal{B}$. The map $F : M \to N$ is a $1-1$ smooth immersion. Then $F(M)$ is a smooth submanifold of N with co-dimension $m - n$.

Solution: 64 Let us set $\tau = \{F(U) : U$ is open in $M\}$. It is obvious that τ is a topology over $F(M)$, and F is a homeomorphism from M onto $F(M)$. Since M is an n-dimensional topological manifold and M is homeomorphic onto $F(M)$, therefore $F(M)$ is an n-dimensional topological manifold. Moreover, $\{(F(U), \phi \circ (F^{-1}|_{F(U)})) : (U, \phi) \in \mathcal{A}\}$ forms an C^∞-atlas on $F(M)$. Hence, $F(M)$ forms a smooth manifold, and F forms a diffeomorphism from M onto $F(M)$.

Now we want to prove that the map $i : F(M) \hookrightarrow N$ is a smooth map. Here $i = F \circ F^{-1}$. Since $F : M \to F(M)$ is a diffeomorphism, $F^{-1} : F(M) \to M$ is also so. Furthermore, as F is a $1 - 1$ smooth immersion, the composition map $i = F \circ F^{-1}$ is a smooth immersion. It follows that $F(M)$ is a smooth submanifold of N with co-dimension $m - n$.

Problem 2.65 Let M be an n-dimensional smooth manifold and $S(\neq \phi) \subset M$. Let S be a smooth submanifold of M with co-dimension 0. Prove that S is an embedded submanifold of M with co-dimension 0.

Solution: 65 Since S is a smooth submanifold of M with co-dimension 0, by its definition, there exists a topology τ over S with respect to which S becomes an n-dimensional topological manifold, and there exists a C^∞ structure on S with respect to which the topological manifold S becomes an n-dimensional smooth manifold such that the map $i : S \hookrightarrow M$ is a $1 - 1$ smooth immersion. Thus i is a smooth embedding, hence a topological embedding. Thus i is a homeomorphism from S onto $i(S)(= S)$, where $i(S)$ has the subspace topology inherited from M. It follows that τ is the subspace topology of S inherited from M. Thus combining all the facts, we conclude that S is an embedded submanifold of M with co-dimension 0.

Problem 2.66 Let M be an n-dimensional smooth manifold and $S(\neq \phi) \subset M$. Let S be a smooth submanifold of M with co-dimension k. If the map $i : S \hookrightarrow M$ is proper, then prove that S is an embedded submanifold of M with co-dimension k.

Solution: 66 Since S is a smooth submanifold of M with co-dimension k, by virtue of its definition, $\exists$ a topology τ over S with respect to which S becomes an $(n - k)$-dimensional topological manifold, and $\exists$ a smooth structure on S with respect to which the topological manifold S becomes an $(n - k)$-dimensional smooth manifold with the inclusion map $i : S \hookrightarrow M$ which is a $1 - 1$ smooth immersion. Since the map i is proper, it follows that i is a smooth embedding (refer to Lemma 2.5), and hence a topological embedding. So i is a homeomorphism from S onto $i(S)(= S)$, where $i(S)$ has the subspace topology inherited from M. It follows that τ is the subspace topology of S inherited from M. Thus combining all the facts, we conclude that S is an embedded submanifold of M with co-dimension k.

Problem 2.67 Let M be an n-dimensional smooth manifold and $S(\neq \phi) \subset M$. Let S be a compact smooth submanifold of M with co-dimension k. Prove that S is an embedded submanifold of M with co-dimension k.

Solution: 67 Since S is a smooth submanifold of M with co-dimension k, by virtue of its definition, $\exists$ a topology τ over S with respect to which S becomes an $(n - k)$-dimensional topological manifold, and $\exists$ a smooth structure on S with respect to which the topological manifold S becomes an $(n - k)$-dimensional smooth manifold with the map $i : S \hookrightarrow M$ which is a $1 - 1$ smooth immersion. Further, since S is compact, it follows that i is a smooth embedding (refer to Lemma 2.5), and hence a topological embedding. So i is a homeomorphism from S onto $i(S)(= S)$, where $i(S)$ has the subspace topology inherited from M. It follows that τ is the subspace topology of S inherited from M. Thus combining all the facts, we conclude that S is an embedded submanifold of M with co-dimension k.

Remark 2.31 Let M be an n-dimensional smooth manifold and N be an m-dimensional smooth manifold. The map $F : M \to N$ is a smooth map. Let $S(\neq \phi) \subset M$ be a smooth submanifold of M. Then the restricted map $F|_s : S \to M$ is also smooth.

Problem 2.68 Let M be an n-dimensional smooth manifold and N be an m-dimensional smooth manifold with $S(\neq \phi) \subset N$. Let S be a smooth submanifold of N with co-dimension k. Let $F : M \to N$ be a smooth map, and $F(M) \subset S$. Let $F : M \to S$ be continuous. Then prove that $F : M \to S$ is a smooth map.

Solution: 68 Let us fix any $p \in M$. Then $F(p) \in F(M) \subset S \Rightarrow F(p) \in S$. Since S is a smooth submanifold of N with co-dimension k, by virtue of its definition, $\exists$ a topology τ over S with respect to which S becomes an $(n - k)$-dimensional topological manifold, and $\exists$ a C^∞ structure on S with respect to which the topological manifold S becomes an $(n - k)$-dimensional smooth manifold with the map $i : S \hookrightarrow N$ which is a $1 - 1$ smooth immersion. Taking advantage of Exercise 2.14, we can say that i is a local diffeomorphism. Hence, $\exists$ an open neighbourhood V of $F(p) \in S$ such that $i(V)$ is open in N, and the map $i|_V : V \to i(V)$ is a diffeomorphism, and hence the map $(i|_v)^{-1} : i(V) \to V$ is smooth. Since V is an open neighbourhood of $F(p) \in S$, and F is continuous, $\exists$ an open neighbourhood U of $p \in M$ such that $F(U) \subset V$. Since U is open in M, U is an open submanifold of M, hence U is an embedded submanifold of M. Thus U is a smooth submanifold of M. Further, since F is a smooth map, by Remark 2.31, $F|_U : U \to N$ is a smooth map. Thus, the composite map $(i|_v)^{-1} \circ F|_U = F|_U : U \to V$ is smooth. Consequently, as U is an open neighbourhood of $p \in M$, $F : M \to S$ is smooth at p, it follows that $F : M \to S$ is a smooth map.

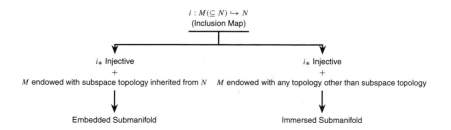

Exercise

Exercise 2.39 *Let M be an n-dimensional smooth manifold, N be an m-dimensional smooth manifold and $S(\neq \phi) \subset N$. Let S be an embedded submanifold of N with co-dimension k. Let $F : M \to N$ be a smooth map, and $F(M) \subset S$. Let $F : M \to S$ be continuous. Then prove that $F : M \to S$ is smooth.*

2.14 f-Related Vector Fields

Let $f : M \to N$ be a smooth map. For each $p \in M$, let $X_p \in T_p(M)$ and $X_{f(p)} \in T_{f(p)}(N)$ be such that

$$f_*(X_p) = Y_{f(p)}. \tag{2.38}$$

In such a case, we say that X, Y are f-related vector fields. Now

$$\{f_*(X_p)\}g = Y_{f(p)}g, \ \forall \ g \in F(f(p)).$$

Using (2.23) and (2.34), we find

$$X_p(g \circ f) = (Yg)(f(p))$$
$$\text{or } \{X(g \circ f)\}(p) = (Yg)f(p) \quad \text{by (2.23)}$$
$$X(g \circ f) = (Yg)f. \tag{2.39}$$

If f is a transformation on M and

$$f_*(X_p) = X_{f(p)}$$
$$i.e. \ (f_*X)_{f(p)} = X_{f(p)},$$

we say that X is f-**related to itself** or X **is invariant under** f. Thus

$$f_*X = X. \tag{2.40}$$

Proposition 2.9 *Let $f : M \to N$ be a smooth map. If the vector fields X_1, X_2 on M are f-related to the vector fields Y_1, Y_2 respectively on N, then the Lie bracket $[X_1, X_2]$ is f-related to the Lie bracket $[Y_1, Y_2]$.*

Proof Given that $X_1(g \circ f) = (Y_1g)f$ and $X_2(g \circ f) = (Y_2g)f$. Now

$$[X_1, X_2](g \circ f) = X_1\{X_2(g \circ f)\} - X_2\{X_1(g \circ f)\}, \quad f \in F(M), g \in F(N)$$
$$= X_1\{(Y_2g)f\} - X_2\{(Y_1g)f\}, \quad \text{from above}$$
$$= \{Y_1(Y_2g)\}f - \{Y_2(Y_1g)\}f, \quad Y_2g \in F(N), \ Y_1g \in F(N)$$
$$= \{Y_1(Y_2g) - Y_2(Y_1g)\}f = \{[Y_1, Y_2]g\}f.$$

Thus by (2.39), we can claim that $[X_1, X_2]$ is *f*-related to $[Y_1, Y_2]$.

Problem 2.69 Let $f : \mathbb{R}^2 \to \mathbb{R}^4$ be a differential map, and $f(x_1, x_2) = (u, v, w, t)$ be such that $u = x_1^2 - x_2^2, v = x_1^2 + x_2^2, w = x_1 + x_2, t = x_1 - x_2$. Let $X = x_1\dfrac{\partial}{\partial x_1} + x_2\dfrac{\partial}{\partial x_2}$ and $Y = -x_2\dfrac{\partial}{\partial x_1} + x_1\dfrac{\partial}{\partial x_2}$ be two vector fields on $\mathbb{R}^2$. Find vector fields on $\mathbb{R}^4$, *f*-related to X, Y respectively.

Solution: 69 By definition $f_* : T_p(\mathbb{R}^2) \to T_{f(p)}(\mathbb{R}^4)$ and if $f_*X = \bar{X}$ and $f_*Y = \bar{Y}$, then $\bar{X}$ and $\bar{Y}$ are respectively the *f*-related vector fields of X, Y. Let

$$\bar{X} = a\frac{\partial}{\partial u} + b\frac{\partial}{\partial v} + c\frac{\partial}{\partial w} + t\frac{\partial}{\partial t}, \tag{2.41}$$

where a, b, c, d are from $F(\mathbb{R}^4)$, to be determined. Now

$$\{f_*X\}u = \bar{X}u = a.$$

Again by (2.34), the left-hand side of the foregoing equation reduces to

$$X(u \circ f) = a.$$

Applying the hypothesis,

$$a = (x_1\frac{\partial}{\partial x_1} + x_2\frac{\partial}{\partial x_2})(x_1^2 - x_2^2) = u(2x_1) - x_2(2x_2) = 2(x_1^2 - x_2^2).$$

Similarly, $\quad b = (x_1\frac{\partial}{\partial x_1} + x_2\frac{\partial}{\partial x_2})(x_1^2 + x_2^2) = x_1(2x_1) + x_2(2x_2) = 2(x_1^2 + x_2^2).$

$$c = (x_1\frac{\partial}{\partial x_1} + x_2\frac{\partial}{\partial x_2})(w \circ f) = (x_1\frac{\partial}{\partial x_1} + x_2\frac{\partial}{\partial x_2})(x_1 + x_2) = x_1 \cdot 1 + x_2 \cdot 1 = x_1 + x_2.$$

$$d = (x_1\frac{\partial}{\partial x_1} + x_2\frac{\partial}{\partial x_2})(t \circ f) = (x_1\frac{\partial}{\partial x_1} + x_2\frac{\partial}{\partial x_2})(x_1 - x_2) = x_1 - x_2.$$

Consequently, from (2.41) we write

$$\bar{X} = 2(x_1^2 - x_2^2)\frac{\partial}{\partial u} + 2(x_1^2 + x_2^2)\frac{\partial}{\partial v} + (x_1 + x_2)\frac{\partial}{\partial u} + (x_1 - x_2)\frac{\partial}{\partial t}$$

$$\text{i.e. } \bar{X} = 2u\frac{\partial}{\partial u} + 2v\frac{\partial}{\partial v} + w\frac{\partial}{\partial w} + t\frac{\partial}{\partial t}$$

is *f*-related vector field of X, in $\mathbb{R}^4$. Again, we write

$$\bar{Y} = a'\frac{\partial}{\partial u} + b'\frac{\partial}{\partial v} + c'\frac{\partial}{\partial \omega} + d'\frac{\partial}{\partial t}, \tag{2.42}$$

where a', b', c', d' are from $F(\mathbb{R}^4)$, to be determined. In a similar manner, it can be shown that

$$\bar{Y} = (t^2 - \omega^2)\frac{\partial}{\partial u} + t\frac{\partial}{\partial \omega} - \omega\frac{\partial}{\partial t}$$

is f-related vector field of Y, in $\mathbb{R}^4$.

Problem 2.70 Let $f : M \to N$ be a C^∞ map. Prove that two vector fields X, Y respectively on M, N are f-related if and only if

$$f^*((f_*X)g) = X(f^*g), \quad g \in F(N).$$

Solution: 70 Given that X, Y are f-related, hence by (2.38) we obtain

$$f_*(X_p) = Y_{f(p)}$$
$$\text{or } (f_*X)_{f(p)} = Y_{f(p)}, \quad \text{by (2.35)}$$
$$\text{or } (f_*X)_{f(p)}g = Y_{f(p)}g, \quad g \in F(N)$$
$$\text{i.e. } \{(f_*X)g\}f(p) = (Yg)f(p), \quad \text{by (2.23)}$$
$$\text{i.e. } (f_*X)g = Yg. \tag{2.43}$$

In view of (2.39), we have

$$X(g \circ f) = (Yg) \circ f$$
$$\text{or } X(f^*g) = f^*(Yg). \quad \text{by (2.32)}$$

Using (2.43) above, on the right-hand side, we get

$$X(f^*g) = f^*((f_*X)g).$$

The converse follows immediately.

Problem 2.71 Let $f : M \to N$ be a C^∞ map. Let X, Y be two f-related vector fields. If σ is the integral curve of X, prove that $f \circ \sigma$ is the integral curve of Y.

Solution: 71 If σ is the integral curve of X, then by definition, we have

$$X_p = X_{\sigma(t_0)},$$

where $\sigma : [a, b] \subset \mathbb{R} \to M$ is such that $\sigma(t_0) = p$, $p \in M$, $t \in [a, b]$ and X_p is tangent vector to the curve σ at p, i.e.

$$X_pf = \frac{d}{dt}f(\sigma(t))\Big|_{t=t_0}.$$

Again by virtue of (2.34), one gets

$$\{f_*(X_p)\}g = X_p(g \circ f) = \frac{d}{dt}(g \circ f)(\sigma(t)), \quad g \in F(N).$$

Using (2.38) on the left-hand side of the last equation, we get

$$\{Y_{f(p)}\}g = \frac{d}{dt}g((f \circ \sigma)(t))\Big|_{t=t_0}.$$

Thus, we can say that $Y_{f(p)}$ is tangent vector to the curve $f \circ \sigma$ at $f(p) = (f \circ \sigma)(t_0)$. Thus, $f \circ \sigma$ is the integral curve of the vector field Y at $f(p)$, i.e.

$$Y_{f(p)} = Y_{f(\sigma(t_0))},$$

where X, Y are f-related vector fields satisfying (2.38).

Problem 2.72 Let the projection map $\pi : \mathbb{R}^2 \to \mathbb{R}$ be defined by $\pi(x, y) = x$. Find the condition that a vector field of $\mathbb{R}^2$ is π-related to some vector field of $\mathbb{R}$.

Solution: 72 Let $X \in \chi(\mathbb{R}^2)$ be such that $X = \xi\dfrac{\partial}{\partial x} + \eta\dfrac{\partial}{\partial y}$ where $\xi : \mathbb{R}^2 \to \mathbb{R}, \eta :$ $\mathbb{R}^2 \to \mathbb{R}$ are c^∞ functions.

Let X be π-related to vector field $Y \in \chi(\mathbb{R})$ and hence by (2.38), we can write

$$\pi_* X = Y. \tag{2.44}$$

Let us write $Y = \theta\dfrac{d}{dt}$, where t denotes the canonical coordinates on $\mathbb{R}$. Again $\pi(x, y) = x$, so we have

$$(\pi_*) = \left(\frac{\partial \pi_1}{\partial x} \quad \frac{\partial \pi_1}{\partial y}\right) = (1 \quad 0) \quad \text{and}$$

$$(\pi_*)(X) = (1 \quad 0)\begin{pmatrix} \xi \\ \eta \end{pmatrix} = \xi.$$

If (2.44) holds, then we must have $\xi = \theta$ and this is the required condition.

Exercises

Exercise 2.40 If f is a transformation on M, prove that for every $X \in \chi(M)$, there exists a unique f-related vector field of X.

Exercise 2.41 Let $f : \mathbb{R}^2 \to \mathbb{R}^3$ be a differential map $f(x_1, x_2) = (u, v, w)$ be such that $u = x_1 x_2, v = x_2 + 1, w = x_1 + 1$. Let $X = x_1^2\dfrac{\partial}{\partial x_1} + x_2\dfrac{\partial}{\partial x_2}, Y = x_1\dfrac{\partial}{\partial x_1}$ be two vector fields on $\mathbb{R}^2$. Find vector fields on $\mathbb{R}^3$, f-related to X, Y respectively.

Answer

2.41. $uw\dfrac{\partial}{\partial u} + (v-1)\dfrac{\partial}{\partial v} + (\omega-1)^2\dfrac{\partial}{\partial \omega}; \quad u\dfrac{\partial}{\partial u} + (\omega-1)\dfrac{\partial}{\partial \omega}.$

2.15 One Parameter Group of Transformations on a Manifold

In this section, we wish to interpret the algebraic interpretation of the vector field.

Let a mapping $\phi : \mathbb{R} \times M \to M$ be such that

$$
\begin{cases}
(i) \text{ for each } t \in \mathbb{R}, \ \phi(t, p) \to \phi_t(p) \text{ is a transformation on } M; \\
(ii) \text{ for all } t, s, t + s \text{ in } \mathbb{R}, \\
\qquad \phi_t(\phi_s(p)) = \phi_{t+s}(p).
\end{cases}
\tag{2.45}
$$

Then the family $\{\phi_t | t \in \mathbb{R}\}$ of mappings is called a one-parameter group of transformations on M.

Problem 2.73 Let $\{\phi_t | t \in \mathbb{R}\}$ be one-parameter group of mappings of M. Show that

(i) ϕ_0 is the identity mapping.
(ii) $\phi_{-t} = (\phi_t)^{-1}$.

Solution: 73 **(i)** Taking $t = 0 \in \mathbb{R}$ in (2.45) (ii), one gets $\phi_0(\phi_s(p)) = \phi_s(p)$. Thus ϕ_0 is the identity mapping.
(ii) For every $t, -t \in \mathbb{R}$

$$
\phi_t(\phi_{-t}(p)) = \phi_0(p), \quad \text{by } (2.45)(ii)
$$
$$
= p, \quad \text{by } (2.45)(i)
$$
$$
\text{or } \phi_{-t}(p) = (\phi_t)^{-1}(p).
$$
$$
\therefore \ \phi_{-t} = (\phi_t)^{-1}, \quad \forall \ p \in M.
$$

Exercise

Exercise 2.42 *Prove that $\{\phi_t | t \in \mathbb{R}\}$ forms an Abelian group.*

Remark 2.32 Exercise 2.42 gives the algebraic interpretation of the vector field X on a manifold.

Let us set

$$
\psi(t) = \phi_t(p).
\tag{2.46}
$$

Then $\psi : \mathbb{R} \to M$ is a differentiable curve on M such that $\psi(0) = \phi_0(p) = p$ (refer to Problem 2.73). Such a curve is called the **orbit** through p of M. The tangent vector, say X_p, to the curve $\psi(t)$ at p is therefore

$$X_p f = \frac{d}{dt} f(\psi(t))|_{t=0} = \lim_{t \to 0} \frac{f(\phi_t(p)) - f(\phi_0(p)))}{t}, \quad \forall f \in F(M). \qquad (2.47)$$

In this case, we say that $\{\phi_t | t \in \mathbb{R}\}$ induces the vector field X and X is called the **generator** of ϕ_t. The curve $\psi(t)$ defined by (2.46) is called the integral curve of X.

Problem 2.74 Show that the mapping $\phi : \mathbb{R} \times \mathbb{R}^3 \to \mathbb{R}^3$ defined by

$$\phi(t, p) = (p^1 + t, p^2 + t, p^3 + t)$$

is a one-parameter group of transformations on $\mathbb{R}^3$ and the generator is given by

$$\frac{\partial}{\partial x^1} + \frac{\partial}{\partial x^2} + \frac{\partial}{\partial x^3} \text{ where } p = (x^1, x^2, x^3) \in \mathbb{R}^3.$$

Solution: 74 Clearly, ϕ is a transformation on $\mathbb{R}^3$ and

$$\phi_t(\phi_s(p)) = \phi_t(p^1 + s, p^2 + s, p^3 + s); \text{ as defined}$$
$$= \phi_{t+s}(p).$$

Thus, $\{\phi_t | t \in \mathbb{R}\}$ is a one-parameter group of transformations on $\mathbb{R}^3$. Again,

$$X_p = \lim_{t \to 0} \frac{\phi_t(p) - \phi_0(p)}{t} = (1, 1, 1).$$

Now $\{\frac{\partial}{\partial x^1}, \frac{\partial}{\partial x^2}, \frac{\partial}{\partial x^3}\}$ is a basis of $T_p(\mathbb{R}^3)$ and hence $X_p \in T_p(\mathbb{R}^3)$ is given by

$$\frac{\partial}{\partial x^1} + \frac{\partial}{\partial x^2} + \frac{\partial}{\partial x^3}$$

which is the generator of $\{\phi_t\}$.

Problem 2.75 Let $\phi : \mathbb{R} \times \mathbb{R}^2 \to \mathbb{R}^2$ be defined by $\phi_t(p) = (x \cos t - y \sin t, x \sin t + y \cos t)$.

  (i) Show that $\{\phi_t | t \in \mathbb{R}\}$ defines a one-parameter group of transformations on $\mathbb{R}^2$.
 (ii) Find its generator.
(iii) Describe the orbit.
(iv) Prove that X is invariant under ϕ_t, i.e. $(\phi_t)_* X_p = X_{\phi_t(p)}$.

Solution: 75 (i) Note that $|J| = \begin{vmatrix} \cos t & -\sin t \\ \sin t & \cos t \end{vmatrix} = 1 \neq 0$. Hence, ϕ^{-1} exists and using Problem 2.73 (ii), we have

$$\phi_t^{-1}(x', y') = \phi_{-t}(x', y') = (x' \cos t + y' \sin t, -x' \sin t + y' \cos t).$$

It can be shown that

$$\phi_t(\phi_{-t}(x',y')) = (x',y') \quad \text{and} \quad \phi_{-t}(\phi_t(x,y)) = (x,y),$$

and hence we claim that $\{\phi_t\}$ is a transformation of $\mathbb{R}^2$. Finally

$$\phi_t(\phi_s(x,y)) = (x\cos(t+s) - y\sin(t+s), x\sin(t+s) + y\cos(t+s))$$
$$= \phi_{t+s}(x,y).$$

Thus, $\{\phi_t | t \in \mathbb{R}\}$ is a one-parameter group of transformations on $\mathbb{R}^2$.

(ii) From the definition,

$$X_p = \frac{d}{dt}\phi_t(p)\Big|_{t=0} = (-x\sin t - y\cos t, x\cos t - y\sin t)\Big|_{t=0} = (-y, x).$$

Thus, the generator is given by $-y\dfrac{\partial}{\partial x} + x\dfrac{\partial}{\partial y}$.

(iii) The orbit through $p = (x_0, y_0)$ while $t = 0$ is the image of the map $\mathbb{R} \to \mathbb{R}^2$ given by

$$t \mapsto (x_0\cos t - y_0\sin t, x_0\sin t + y_0\cos t).$$

(iv) Again, $X = -y\dfrac{\partial}{\partial x} + x\dfrac{\partial}{\partial y}$ (refer to (ii) above). Now

$$\phi_t(p) = \phi_t(x_0, y_0) = (x_0\cos t - y_0\sin t, x_0\sin t + y_0\cos t).$$

Thus

$$X_{\phi_t(p)} = \left(-y\frac{\partial}{\partial x} + x\frac{\partial}{\partial y}\right)_{(x_0\cos t - y_0\sin t,\, x_0\sin t + y_0\cos t)}$$
$$= (-x_0\sin t - y_0\cos t)\frac{\partial}{\partial x} + (x_0\cos t - y_0\sin t)\frac{\partial}{\partial y}.$$

Again

$$(\phi_t)_* X_p = \begin{pmatrix} \cos t & -\sin t \\ \sin t & \cos t \end{pmatrix}\begin{pmatrix} -y \\ x \end{pmatrix}\Bigg|_p$$
$$= (-y\cos t - x\sin t, -y\sin t + x\cos t)_{(x_0, y_0)}$$
$$= (-y_0\cos t - x_0\sin t, -y_0\sin t + x_0\cos t).$$

Thus

$$(\phi_t)_* X_p = (-x_0\sin t - y_0\cos t)\frac{\partial}{\partial x} + (x_0\cos t - y_0\sin t)\frac{\partial}{\partial y} = X_{\phi_t(p)}.$$

Thus X is invariant under ϕ_t.

Problem 2.76 Let $\phi : \mathbb{R} \times \mathbb{R}^2 \to \mathbb{R}^2$ be defined by $\phi_t(x, y) = (x, ye^t)$.

(i) Show that $\{\phi_t | t \in \mathbb{R}\}$ defines a one-parameter group of transformations on $\mathbb{R}^2$.
(ii) Find its generator.
(iii) Describe the orbit.
(iv) Prove that X is invariant under ϕ_t.

Solution: **(i)** Note that $|J| = \begin{vmatrix} 1 & 0 \\ 0 & e^t \end{vmatrix} = e^t \neq 0, \forall t$. Hence ϕ^{-1} exists and

$$\phi_t^{-1}(x', y') = \phi_{-t}(x', y') = (x', y'/e^t).$$

It can be shown that

$$\phi_t(\phi_{-t}(x', y')) = (x', y') \quad \text{and} \quad \phi_{-t}(\phi_t(x, y)) = (x, y),$$

and hence we claim that $\{\phi_t\}$ is a transformation of $\mathbb{R}^2$. Finally

$$\phi_t(\phi_s(x, y)) = (x, ye^{t+s})$$
$$= \phi_{t+s}(x, y).$$

Thus, $\{\phi_t | t \in \mathbb{R}\}$ is a one-parameter group of transformations on $\mathbb{R}^2$.
(ii) From the definition,

$$X_p = \frac{d}{dt}\phi_t(p)\Big|_{t=0} = (0, ye^t)\Big|_{t=0} = (0, y).$$

Thus the generator is given by $y\dfrac{\partial}{\partial y}$.
(iii) The orbit through $p = (x_0, y_0)$ while $t = 0$ is the image of the map $\mathbb{R} \to \mathbb{R}^2$ given by

$$t \mapsto (x_0, y_0 e^t).$$

(iv) Now $\phi_t(p) = \phi_t(x_0, y_0) = (x_0, y_0 e^t)$. Hence $X_{\phi(t)(p)} = \left(y\dfrac{\partial}{\partial y}\right)_{(x_0, y_0 e^t)} = \left(y_0 e^t\right)\dfrac{\partial}{\partial y}$.
Again

$$(\phi_t)_* X_p = \begin{pmatrix} 1 & 0 \\ 0 & e^t \end{pmatrix}\begin{pmatrix} 0 \\ y \end{pmatrix}\Bigg|_{(x_0, y_0)}$$

$$= (0, ye^t)_{(x_0, y_0)}$$

$$= \left(y_0 e^t\right)\frac{\partial}{\partial y} = X_{(\phi_t)(p)}.$$

Thus X is invariant under ϕ_t.

Exercises

Exercise 2.43 *Show that the following families of maps* $\phi : \mathbb{R} \times \mathbb{R}^2 \to \mathbb{R}^2$ *form a one-parameter group of transformations and find their generators.*

(i) $\phi_t(p) = (x + at, y + bt), \ a, b \in \mathbb{R}$
(ii) $\phi_t(p) = (xe^{2t}, ye^{-2t})$

where $p = (x, y)$.

Exercise 2.44 *Let* $\phi : \mathbb{R} \times \mathbb{R}^2 \to \mathbb{R}^2$ *be defined by*

$$\phi_t(p) = (x \cos t + y \sin t, -x \sin t + y \cos t), \ \ p = (x, y).$$

Show that $\{\phi_t | t \in \mathbb{R}\}$ *defines a one-parameter group of transformations on* $\mathbb{R}^2$.
(i) Find its generator X.
(ii) Describe the orbit.
(iii) Prove that X *is invariant under* ϕ_t, *i.e.* $(\phi_t)_* X_p = X_{\phi_t(p)}$.

Exercise 2.45 *Let* $M = GL(2, \mathbb{R})$ *and a mapping* $\phi_t(A) = \begin{pmatrix} 1 & t \\ 0 & 1 \end{pmatrix} \cdot A$, $A \in GL(2, \mathbb{R})$ *with the dot denoting matrix multiplication. Find the generator.*

Answers

2.43 (i) $a \dfrac{\partial}{\partial x} + b \dfrac{\partial}{\partial y}$ (ii) $2x \dfrac{\partial}{\partial x} - 2y \dfrac{\partial}{\partial y}$.

2.44 (i) $y \dfrac{\partial}{\partial x} - x \dfrac{\partial}{\partial y}$ (ii) circle centred at the origin.

2.45 $a_{21} \dfrac{\partial}{\partial x} + a_{22} \dfrac{\partial}{\partial y}$.

Problem 2.77 Let $M = \mathbb{R}^3$ and a mapping $\phi : \mathbb{R} \times M \to M$ be such that $X = \dfrac{\partial}{\partial x}$ is its generator. Find ϕ.

Solution: 76 For X, the differential equations are

$$\frac{dx}{dt} = 1 \quad \frac{dy}{dt} = 0 = \frac{dz}{dt}, \quad \text{where } (x, y, z) \in \mathbb{R}^3.$$

On solving, we get
$$x = t + A, \quad y = B, \quad z = C$$

where A, B, C are integrating constants. If for $t = 0, x = x_0, y = y_0, z = z_0$, then $A = x_0, B = y_0, C = z_0$. Consequently, the integral curve is given by $\psi(t) = (x_0 + t, y_0, z_0)$ and $\phi_t : \mathbb{R} \times \mathbb{R}^3 \to \mathbb{R}^3$ is defined by $\phi_t(x, y, z) = (x + t, y, z)$ where $\phi_0(x, y, z) = \psi(0) = p = (x_0, y_0, z_0)$.
In this case

$$\phi_t(p) = \phi_t(x_0, y_0, z_0) = (x_0 + t, y_0, z_0),$$

and hence

$$X_{\phi_t(p)} = \left(\frac{\partial}{\partial x}\right)_{(x_0+t, y_0, z_0)} = (x_0 + t)\frac{\partial}{\partial x}$$

and

$$(\phi_t)_* X_p = \begin{pmatrix} 1 & 0 & 0 \\ 0 & 1 & 0 \\ 0 & 0 & 1 \end{pmatrix} \begin{pmatrix} 1 \\ 0 \\ 0 \end{pmatrix}\Bigg|_p = (1, 0, 0)_{(x_0+t, y_0, z_0)} = (x_0 + t, 0, 0)$$

$$= (x_0 + t, 0, 0)\frac{\partial}{\partial x} = X_{(\phi_t)(p)}.$$

Exercise 2.46 *Let $M = \mathbb{R}^2$, the xy-plane and $X = y\dfrac{\partial}{\partial x} - x\dfrac{\partial}{\partial y}$. Find the domain W and the one-parameter group $\phi : W \to M$.*

Exercise 2.47 *Let $M = \mathbb{R}^2$ and a mapping $\phi : \mathbb{R} \times M \to M$ be such that*

(i) $X = x\dfrac{\partial}{\partial x} + y\dfrac{\partial}{\partial y}$ *is its generator;*

(ii) $X = -y\dfrac{\partial}{\partial x} + x\dfrac{\partial}{\partial y}$ *is its generator;*

(iii) $X = \dfrac{\partial}{\partial x} + y\dfrac{\partial}{\partial y}$ *is its generator;*

Find ϕ in each case.

Answers
2.46 $(yt + a, -xt + b)$.
(2.47)(i) $\phi_t(x, y) = (xe^t, ye^t)$ (ii) $\phi_t(x, y) = (x\cos t - y\sin t, x\sin t + y\cos t)$.
  (iii) $\phi_t(x, y) = (t + x, ye^t)$.
  Since every one-parameter group of transformations generates a vector field, the question now arises whether **every** vector field induces a one-parameter group of transformations or not. The question has been answered in **negative**.

Example 2.23 Let $X = -e^x\dfrac{\partial}{\partial x} + \dfrac{\partial}{\partial y}$ be defined on $\mathbb{R}^2$. As done earlier, it can be shown that the integral curve $\psi(t)$ of X is $\psi(t) = \left(\log\dfrac{1}{(t + e^{-p^1})}, t + p^2\right)$, **not defined** $\forall t \in \mathbb{R}$, where $x(0) = p^1, y(0) = p^2$.
  Consequently, by (2.46), if we define $\psi(t) = \phi_t(p)$ then, X does not induce one-parameter group of transformations on $\mathbb{R}^2$.

  The above observation leads to the following definition:
Local one-**parameter group of transformations:** Let I_ϵ be an open interval $(-\epsilon, \epsilon)$ on $\mathbb{R}$ and U be a neighbourhood of a point p of M (Fig. 2.21).

Fig. 2.21 Local
one-parameter group of
transformations

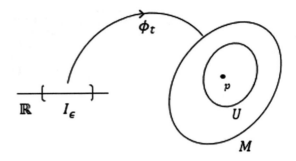

Let us define a mapping $\phi : I_\epsilon \times U \to \phi_t(U) \subset M$ by $\phi(t, p) \mapsto \phi_t(p)$ be such that

(i) U is an open cover of M.
(ii) for each $t \in I_\epsilon$, $\phi_t(p)$ is a transformation of U onto an open subset $\phi_t(U)$ of M.
(iii) if $t, s, t + s$ are in I_ϵ and if $\phi_s(p) \in U$ then

$$\phi_t(\phi_s(p)) = \phi_{t+s}(p).$$

Such a family $\{\phi_t | t \in I\}$ of mappings is called a **local one-parameter group of transformations defined on** $I_\epsilon \times U$.

Now, we are going to prove the following theorem.

Theorem 2.7 *Let X be a vector field on a manifold M. Then X generates local one-parameter group of transformations in a neighbourhood of a point in M.*

Proof Let (U, ϕ) be a chart of p of M. By Exercise 2.8 we can write

$$\phi(p) = (0, 0, 0, \ldots, 0) \in \mathbb{R}^n.$$

If $(x^1, x^2, \ldots, x^n)$ is the local coordinate system of p, then $x^i(p) = 0$, $i = 1, 2, 3, \ldots, n$ (Fig. 2.22).

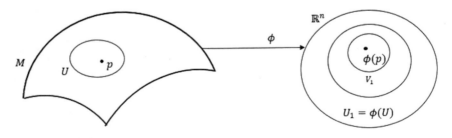

Fig. 2.22 Existence theorem of local one-parameter group of transformations

Let $X = \xi^i \dfrac{\partial}{\partial x^i}$ be a given vector field on U of p with each $\xi^i : U \subset \mathbb{R}^n \to \mathbb{R}$, $i = 1, 2, 3, \ldots, n$ being differentiable. Then ϕX, the ϕ-related vector field on $\mathbb{R}^n$, is defined in a neighbourhood $U_1 = \phi(U)$ at $\phi(p) = (0, 0, 0, 0 \ldots, 0) \in \mathbb{R}^n$. We write $\phi X = \eta^i \dfrac{\partial}{\partial x^i}$ where each $\eta^i : \phi(U) \subset \mathbb{R}^n \to \mathbb{R}$ is differentiable. Then by virtue of the existence theorem of the ordinary differential equation, for each $\phi(p) \in U_1 \subset \mathbb{R}^n$, there exists $\delta_1 > 0$ and a neighbourhood V_1 of $\phi(p)$, $V_1 \subset U_1$ such that, for each $q = (q^1, q^2, \ldots, q^n) \in V_1$, $\phi(r) = q$, say, $r \in U \subset M$, there exists n-tuple of C^∞ functions $f^1(t, q), f^2(t, q), \ldots, f^n(t, q)$ defined on $I_{\delta_1} \subset I_{\epsilon_1}$,

$$f^i : I_{\delta_1} \to V_1 \subset U_1 \subset \mathbb{R}^n, \ i = 1, 2, 3, \ldots, n$$

which satisfies the system of first-order differential equations

$$\frac{d}{dt} f^i(t) = \eta^i(t, \phi(p)), \ i = 1, 2, 3, \ldots, n \tag{2.48}$$

with the initial condition

$$f^i(0, q) = f^i(0) = q^i. \tag{2.49}$$

Let us write

$$\theta_t(q) = (f^1(t, q), f^2(t, q), \ldots, f^n(t, q)). \tag{2.50}$$

Then $\theta : I_{\delta_1} \times V_1 \to \theta_t(q) \in V_1$ is a transformation of V_1 onto an open set $\theta_t(V_1)$ of $\mathbb{R}^n$.

Let us set

$$(g^1(t), g^2(t), \ldots, g^n(t)) = (f^1(t + s, q), f^2(t + s, q), \ldots, f^n(t + s, q)),$$

where each $f^i(t + s, q)$, $f^i(t, s(q))$ are defined on $I_{\delta_1} \times V_1$ if $\theta_s(q) \in V_1 \subset U_1$ and $t, s, t + s$ are in I_{δ_1}. Componentwise, we write

$$(g^i(t)) = (f^i(t + s, q)),$$

where each $g^i(t)$ is defined on $I_{\delta_1} \times V_1$, $V_1 \subset U_1$ with initial condition

$$(g^i(0)) = (f^i(s, q)). \tag{2.51}$$

Similarly, if we write

$$(h^i(t)) = (f^i(t, \theta_s(q))),$$

then each $h^i(t)$ is defined on $I_{\delta_1} \times V_1$, $V_1 \subset U_1$ and hence satisfies (2.48) with initial condition

$$(h^i(0)) = (f^i(0, \theta_s(q))) = (f^i(0)) = (\theta_s(q))^i = (f^i(s, q)), \text{ by } (2.49), (2.50)$$

as constructed. Thus $(h^i(0)) = (g^i(0))$ (by (2.51)). Hence, from uniqueness we have

$$(h^i(t)) = (g^i(t)) \quad i.e. \quad (f^i(t, \theta_s(q))) = (f^i(t + s, q)),$$

which can be written as $\theta_t(\theta_s(q)) = \theta_{t+s}(q)$ (refer to (2.50)). We can also write it
as $\theta(t, \theta(s, q)) = \theta(t + s, q)$. Thus, $\{\theta_t | t \in I_{\delta_1}\}$ is the local one-parameter group of
transformations induced by the vector field ϕX at U_1 of $\phi(p)$ of $\mathbb{R}^n$.

Let us now set $\phi^{-1}(V_1) = V \subset U$ of p of M and define

$$\psi : I_\epsilon \times V \to \psi_t(V) \subset M,$$

as $\psi_t(r) = \phi^{-1}(\theta_t(q))$ with $q = \phi(r)$, i.e.

$$\psi_t(r) = \phi^{-1}(\theta(t, \phi(r))). \tag{2.52}$$

Then

   (i) V is an open cover of M
   (ii) for each $t \in I_\epsilon$, $\psi(t, p) \mapsto \psi_t(p)$ is a transformation of V onto an open set
         $\psi_t(V)$ of M and
   (iii) if $t, s, t + s$ are in I_ϵ and if $\psi_s(r) \subset \psi_t(V)$, then

$$
\begin{aligned}
\psi_t(\psi_s(r)) &= \phi^{-1}(\theta(t, \phi(\psi_s(r)))), \text{ by } (2.52)\\
&= \phi^{-1}(\theta(t, \phi\phi^{-1}(\theta(s, \phi(r))))), \text{ by } (2.52)\\
&= \phi^{-1}(\theta(t, \theta(s, q)))\\
&= \phi^{-1}(\theta(t + s, q)), \text{ as } \{\theta_t\} \text{ is the local 1-parameter group of transformations}\\
&= \psi_{t+s}(r), \text{ by } (2.52),
\end{aligned}
$$

i.e. $\{\psi_t | t \in I_\epsilon\}$ is the local one-parameter group of transformations defined on $I_\epsilon \times V$, $V \subset U \subset M$ for the vector field X defined in the neighbourhood U of a point of
M.

Finally, if we write

$$
\begin{aligned}
\gamma(t) = \psi_t(r) &= \phi^{-1}(\theta_t(q)), \quad q = \phi(r)\\
&= \phi^{-1}(\sigma(t)), \text{ say}
\end{aligned}
$$

then $\phi^{-1}(\sigma(t))$ is the integral curve of X, where $\sigma(t)$ is the integral curve of the
vector field ϕX of $\mathbb{R}^n$. This completes the proof.

Problem 2.78 Prove that the integral curve always gives rise to a vector field, but
the converse is not true.

Solution: 77 Let $\{\phi_t | t \in \mathbb{R}\}$ be one-parameter group of transformations on a manifold M. From (2.46), we see that if $\psi : \mathbb{R} \to M$ is a differentiable curve on M such that

$$\psi(t) = \phi_t(p), \quad \psi(o) = \phi_0(p) = p,$$

then $X_p = \dfrac{d}{dt}(\psi(t))\big|_{t=0} = \dfrac{d}{dt}(\phi_t(p))\big|_{t=0}$ is called the generator of $\{\phi_t | t \in \mathbb{R}\}$ and the curve $\psi(t)$ is the integral curve of X.

Thus, every one-parameter group of transformations or the integral curve on a manifold induces a vector field on a manifold.

Conversely, by Example 2.23, the vector field on a manifold does not in general induce an integral curve on a manifold.

Problem 2.79 Let ϕ be a transformation on M. If a vector field X generates $\{\phi_t | t \in I_\epsilon\}$ as its local one-parameter group of transformations, prove that the vector field $\phi_* X$ will generate $\{\phi\phi_t\phi^{-1} | t \in I_\epsilon\}$ as its local one-parameter group of transformations on M.

Solution: 78 Let X be given vector field on a manifold M. Then by Theorem 2.7, X generates $\{\phi_t | t \in I_\epsilon\}$ as its local one-parameter group of transformations on M.

Let $\psi(t) = \phi_t(p)$, then X_p is the tangent vector to the curve $\psi(t)$ at $\psi(0) = \phi_0(p) = p$ (refer to Problem (2.73)(i)). Thus

$$X_p = \frac{d}{dt}\psi(t)\big|_{t=0} = \frac{d}{dt}(\phi_t(p))\big|_{t=0}.$$

Now by definition, $\phi_* : T_p(M) \to T_{\phi(p)}(M)$ and $\phi_*(X_p)$ is defined to be the tangent vector to the curve $\phi(\psi(t))$ at $\phi(\psi(0)) = \phi(p)$, i.e.

$$\phi(\psi(t)) = \phi(\phi_t(p))$$
$$= \phi(\phi_t(\phi^{-1}(q))), \text{ say where } \phi(p) = q, \ \phi \text{ being a transformation on } M$$
$$= (\phi\phi_t\phi^{-1})(\phi(p)), \text{ and}$$
$$\phi_*(X_p) = \frac{d}{dt}\phi(\psi(t))\big|_{t=0}$$

or $(\phi_* X)_{\phi(p)} = \dfrac{d}{dt}(\phi\phi_t\phi^{-1})(\phi(p))\big|_{t=0}$, from above and by (2.35).

Comparing with (2.47), we can now say that the vector field $\phi_* X$ generates $\{\phi\phi_t\phi^{-1} | t \in I_\epsilon\}$ as its local one-parameter group of transformations on M.

Exercise

Exercise 2.48 *Show that a vector field X on a manifold M is invariant under a transformation ϕ on M if and only if $\phi \circ \phi_t = \phi_t \circ \phi$ where $\{\phi_t | t \in I\}$ is the local one-parameter group of transformations on M, generated by X.*

Now, we are going to give the geometrical interpretation of the Lie Bracket $[X, Y]$ for every vector field X, Y on M.

Theorem 2.8 *If X generates $\{\phi_t | t \in I\}$ as its local one-parameter group of transformations, then for every vector field Y on M*

$$[X, Y] = \lim_{t \to 0} \frac{1}{t} \{Y - (\phi_t)_* Y\}.$$

We also write

$$[X, Y]_q = \lim_{t \to 0} \frac{1}{t} \{Y_q - ((\phi_t)_* Y)_q\}, \, \forall \, q \in M, q = \phi_t(p), p \in M.$$

To prove the theorem, we require a few lemmas.

Lemma 2.7 *If $\psi(t, p)$ is a function on $I_\epsilon \times M$, $I_\epsilon = (-\epsilon, \epsilon)$ on $\mathbb{R}$ such that $\psi(0, p) = 0$, $\forall \, p \in M$, then there exists a function $h(t, p)$ on $I_\epsilon \times M$ such that $th(t, p) = \psi(t, p)$. Moreover*

$$h(0, p) = \psi'(0, p), \, \psi' = \frac{d\psi}{dt}.$$

Proof Let us define $h(t, p) = \displaystyle\int_0^1 \psi'(ts, p) \frac{d(ts)}{t}$. In view of the fundamental theorem of calculus, we have

$$h(t, p) = \frac{1}{t} [\psi(ts, p)]_0^1 \implies t\,h(t, p) = \psi(t, p).$$

Also, $h(0, p) = \displaystyle\int_0^1 \psi'(0, p) ds = \psi'(0, p)$.

Lemma 2.8 *If f is a function on M and X is a vector field on M which induces a local one-parameter group of transformations $\{\phi_t | t \in I\}$, then there exists a function g_t defined on $I_\epsilon \times V$, V being the neighbourhood of a point p of M where $g_t(p) = g(t, p)$ such that $f(\phi_t(p)) = f(p) + t\,g_t(p)$. Moreover, $X_p f = g(0, p) = g_0(p)$. Symbolically, $Xf = g_0$.*

Proof Let us set

$$\tilde{f}(t, p) = f(\phi_t(p)) - f(\phi_0(p)), \, \forall \, p \in M.$$

Then $\tilde{f}(t, p)$ is a function on $I_\epsilon \times M$ such that $\tilde{f}(0, p) = 0$, $\forall \, p \in M$. Hence by Lemma 2.7, there exists a function, say, $g(t, p)$ on $I_\epsilon \times V$, $V \subset M$ such that

$$t\,g(t, p) = \tilde{f}(t, p)$$
$$\text{or } t\,g_t(p) = f(\phi_t(p)) - f(p).$$

Hence, the result follows. Also from above

$$g(0, p) = \lim_{t \to 0} \frac{1}{t} \{f(\phi_t(p)) - f(\phi_o(p))\} = X_p f.$$

Proof of the Main theorem:

Let us write $\phi_t(p) = q$. Therefore, $p = \phi_{-t}(q)$. Then by (2.34), we get

$$\{((\phi_t)_* Y) f\}(q) = \{Y(f \circ \phi_t)\}(p) = (Yf + t Y g_t)(p), \text{ by Lemma 2.8.}$$

Therefore,

$$(Yf)(q) - \{((\phi_t)_* Y) f\}(q) = (Yf)(q) - (Yf)(p) - t(Y g_t) \phi_{-t}(q)$$

$$\text{or, } \lim_{t \to 0} \frac{1}{t} \{Y_q - ((\phi_t)_* Y)_q\} f = \lim_{t \to 0} \frac{(Yf)(q) - (Yf)(p)}{t} - \lim_{t \to 0} (Y g_t) \phi_{-t}(q)$$

$$= \lim_{t \to 0} \frac{(Yf)(q) - (Yf)(p)}{t} - Y_q(Xf), \text{ by Lemma 2.8.}$$

Now from (2.47),

$$X_q f = \lim_{t \to 0} \frac{1}{t} \{f(\phi_t(q)) - f(q)\}$$

$$\therefore \; -X_q f = \lim_{t \to 0} \frac{1}{t} \{f(p) - f(q)\}, \text{ as } p = \phi_{-t}(q).$$

Replacing f by Yf, one obtains

$$X_q(Yf) = \lim_{t \to 0} \frac{1}{t} \{(Yf)(q) - (Yf)(p)\}.$$

Thus, we write

$$\lim_{t \to 0} \frac{1}{t} \{Y_q - ((\phi_t)_* Y)_q\} f = X_q(Yf) - Y_q(Xf)$$

$$= \{X(Yf) - Y(Xf)\}(q)$$

$$= \{[X, Y] f\}(q)$$

$$= [X, Y]_q f.$$

Therefore, $[X, Y]_q = \lim_{t \to 0} \frac{1}{t} \{Y_q - ((\phi_t)_* Y)_q\}$, $\forall f$. The result follows immediately.

Corollary 2.2 *Show that* $(\phi_s)_*[X, Y] = \lim_{t \to 0} \frac{1}{t} \{(\phi_s)_* Y - (\phi_{s+t})_* Y\}.$

Proof In view of Theorem 2.8, one gets

$$(\phi_s)_*[X, Y] = \lim_{t \to 0} \frac{1}{t}(\phi_s)_*\{Y - (\phi_t)_*Y\}$$

$$= \lim_{t \to 0} \frac{1}{t}\{(\phi_s)_*Y - (\phi_s)_*(\phi_t)_*Y)\}, \text{ as } (\phi_s)_* \text{ is linear}$$

$$= \lim_{t \to 0} \frac{1}{t}\{(\phi_s)_*Y - (\phi_s \circ \phi_t)_*Y\}, \text{ by Problem 2.51}$$

$$= \lim_{t \to 0} \frac{1}{t}\{(\phi_s)_*Y - (\phi_{s+t})_*Y\}, \text{ by (2.45)(ii).}$$

Corollary 2.3 *Show that* $(\phi_s)_*[X, Y] = -\dfrac{d(\phi_t)_*Y}{dt}\bigg|_{t=s}$.

Proof Note that

$$\frac{d}{dt}((\phi_t)_*Y)\bigg|_{t=s} = \lim_{h \to 0} \frac{(\phi_{t+h})_*Y - (\phi_t)_*Y}{h}\bigg|_{t=s}$$

$$= \lim_{h \to 0} \frac{(\phi_{s+h})_*Y - (\phi_s)_*Y}{h}$$

$$= -(\phi_s)_*[X, Y], \text{ using Corollary 2.2.}$$

Corollary 2.4 *Let* X, Y *generate* $\{\phi_t\}$ *and* $\{\psi_s\}$ *respectively as its local one-parameter group of transformations. Then* $\phi_t \circ \psi_s = \psi_s \circ \phi_t$, *if and only if* $[X, Y] = 0$.

Proof Let $\phi_t \circ \psi_s = \psi_s \circ \phi_t$. Then from Exercise 2.48, we can say that the vector field Y is invariant under ϕ_t. Consequently, by (2.40), we find $(\phi_t)_*Y = Y$. Hence, taking advantage of Theorem 2.8, we find $[X, Y] = 0$.

Conversely, let $[X, Y] = 0$. Then in view of the foregoing corollary, we have

$$\frac{d}{dt}((\phi_t)_*Y) = 0$$

$$i.e. \ (\phi_t)_*Y = Constant, \ \forall t$$

$$i.e. \ (\phi_t)_*Y = (\phi_0)_*Y = Y.$$

Finally, taking into consideration Exercise 2.48, we must have the desired result.

A vector field X on a manifold is said to be **complete** if it generates a one-parameter group of transformations on M.

Theorem 2.9 *Every vector field on a compact manifold is complete.*

Proof Left to the reader.

Hint

Theorem 2.9: Use Theorem 2.7 and then use the compactness property.

Remark 2.33 If ϕ is a transformation on a compact manifold and the vector field X is complete, then $\phi_* X$ is also so.

Chapter 3
Differential Forms

3.1 Cotangent Space

A mapping $\omega : \chi(M) \to F(M)$ that satisfies

$$\begin{cases} \omega(X + Y) = \omega(X) + \omega(Y) \\ \omega(bX) \quad = b\omega(X), \ \forall \ X, Y \in \chi(M), \ b \in F(M) \end{cases} \tag{3.1}$$

is called a linear mapping over $\mathbb{R}$, where $\chi(M), \ F(M)$ are vector spaces over $\mathbb{R}$.

A linear mapping $\omega : \chi(M) \to F(M)$ denoted by $X \mapsto \omega(X)$ is also called a 1-form on M.

Let $\mathfrak{D}_1(M) = \{\omega, \mu, \dots, \dots \mid \omega : \chi(M) \to F(M)\}$ be the set of all 1-forms on M. Let us define

$$\begin{cases} (\omega + \mu)(X) = \omega(X) + \mu(X) \\ \omega(bX) \quad = b\omega(X) \end{cases} \tag{3.2}$$

It can be shown that $\mathfrak{D}_1(M)$ is a vector space over $\mathbb{R}$, called the dual of $\chi(M)$. We write

$$\{\omega(X)\}(p) = \omega_p(X_p), \ \text{where} \ \omega(X) \in F(M). \tag{3.3}$$

Thus $\omega_p : T_p(M) \to \mathbb{R}$ and hence $\omega_p \in$ dual of $T_p(M)$.

We denote the dual of $T_p(M)$ by $T_p^*(M)$ and is called the **cotangent space** of M at $p \in M$. Elements of $T_p^*(M)$ are also called the **co-vectors** at $p \in M$.

For every $f \in F(M)$, we denote the total differential of f by df and is defined as

$$(df)_p(X_p) = (Xf)(p) = X_p f, \ \forall \ p \in M. \tag{3.4}$$

We also write it as

$$(df)(X) = Xf. \tag{3.5}$$

© The Author(s), under exclusive license to Springer Nature Singapore Pte Ltd. 2023   129
M. Majumdar and A. Bhattacharyya, *An Introduction to Smooth Manifolds*,
https://doi.org/10.1007/978-981-99-0565-2_3

Problem 3.1 *Show that for every* $f \in F(M)$, df *is a* 1-*form on* M.

Solution: For every $X, Y \in \chi(M)$, $X + Y \in \chi(M)$ and

$$
\begin{aligned}
df(X + Y) &= (X + Y)f, \text{ by } (3.5) \\
&= df(X) + df(Y), \text{ by } (3.5).
\end{aligned}
$$
$$
\begin{aligned}
\text{Also } df(bX) &= (bX)f, \text{ by } (3.5) \\
&= b(Xf) \\
&= bdf(X), \text{ by } (3.5)
\end{aligned}
$$

Thus df is a 1-form.

Exercise

Exercise 3.1 *If* $(x^1, x^2, \ldots, x^n)$ *are co-ordinate functions defined in a neighbourhood of* p *of* M, *show that each* dx^i, $i = 1, 2, 3, 4, \ldots, n$ *is a* 1-*form on* M.

From Exercise 3.1, we claim that each $dx^i \in T_p^*(M)$, $i = 1, 2, 3, \ldots, n$. We define

$$
(dx^i)_p \left(\frac{\partial}{\partial x^j} \right)_p = \delta^i_j = \begin{cases} 1, \, i = j \\ 0, \, i \neq j \end{cases} \tag{3.6}
$$

Let $\omega_p \in T_p^*(M)$ be such that

$$
\omega_p \left(\frac{\partial}{\partial x^j} \right)_p = (f_j)_p, \text{ where each } (f_j)_p \in \mathbb{R}. \tag{3.7}
$$

If possible, let $\mu_p \in T_p^*(M)$ be such that

$$
\mu_p = (f_1)_p (dx^1)_p + (f_2)_p (dx^2)_p + \cdots + (f_n)_p (dx^n)_p.
$$

Then, $\mu_p \left(\dfrac{\partial}{\partial x^i} \right)_p = (f_i)_p, i = 1, 2, \ldots, n$ by (3.6)

$$
= \omega_p \left(\frac{\partial}{\partial x^i} \right)_p, , i = 1, 2, \ldots, n \text{ see } (3.7).
$$

and hence $\mu_p = \omega_p$ as $\left\{ \dfrac{\partial}{\partial x^i} : i = 1, 2, 3, 4, \ldots, n \right\}$ is a basis of $T_p(M)$. Thus any $\omega_p \in T_p^*(M)$ can be expressed uniquely as

$$\omega_p = \sum_{i=1}^{n} (f_i)_p (dx^i)_p, \ i.e. \ \omega = \sum f_i dx^i \tag{3.8}$$

and hence $T_p^*(M) = \text{span}\{(dx^1)_p, (dx^2)_p, \ldots, (dx^n)_p\}$. Finally, if $(f_i)_p (dx^i)_p = 0$
then $(f_i)_p (dx^i)_p \left(\dfrac{\partial}{\partial x^k} \right)_p = 0$ yields $(f_k)_p = 0$ [refer to (3.6)].

Similarly, it can be shown that $(f_1)_p = (f_2)_p = (f_3)_p = \cdots = (f_n)_p = 0$ and the
set $\{(dx^1)_p, (dx^2)_p, \ldots, (dx^n)_p\}$ is linearly independent. We state

Theorem 3.1 *If* $(x^1, x^2, \ldots, x^n)$ *are local co-ordinate system in a neighbourhood*
U of p of M, then the set $\{(dx^1)_p, (dx^2)_p, \ldots, (dx^n)_p\}$ *is a basis of* $T_p^*(M)$ *or*
$\mathfrak{D}_1(M^n)$.

Remark 3.1 A zero-form is nothing but a function, by convention.

Remark 3.2 We say that the form ω in (3.8) is differentiable if each f_i is of
class C^∞.

Remark 3.3 From $\omega = \sum f_i dx^i$, $X = \sum \xi^i \dfrac{\partial}{\partial x^j}$, we see that $\omega(X) = \sum f_i \xi^i$
[refer to (3.6)].

Remark 3.4 $\mathfrak{D}_1(M)$ is a $F(M)$-module.

Unless Otherwise Stated, by a form, we will mean **Differential form**. From
(3.5), we see that

$$dx^i(X) = Xx^i = \sum \xi^j \frac{\partial}{\partial x^j} x^i, \ \text{say}$$

Thus

$$dx^i(X) = \xi^i. \tag{3.9}$$

Consequently, from (3.5), we have

$$df(X) = Xf = \sum \xi^j \frac{\partial}{\partial x^j} f$$
$$= \sum \frac{\partial f}{\partial x^i} dx^i(X), \ \text{by (3.9)}$$
$$df = \sum \frac{\partial f}{\partial x^i} dx^i. \tag{3.10}$$

The transition formula for a 1-form: Let $(U, x^1, x^2, \ldots, x^n)$ and $(V, y^1, y^2, \ldots, y^n)$ be two charts on M where $U \cap V \neq \phi$. Now from (3.8), we have

$$\omega = \sum_i f_i dx^i = \sum_j g_j dy^j, \quad \text{say, where each } f_i, g_i \in F(M)$$

$$= \sum_i \sum_j g_j \frac{\partial y^j}{\partial x^i} dx^i \quad [\text{refer to } (3.10)]$$

$$f_i = \sum_j g_j \frac{\partial y^j}{\partial x^i}, \quad i = 1, 2, 3, 4, \ldots, n \qquad (3.11)$$

as $\{dx^i : i = 1, 2, 3, 4, \ldots, n\}$ is a basis.

For every $f \in F(M)$, $\omega \in \mathfrak{D}_1(M)$, we define $f\omega \in \mathfrak{D}_1(M)$ as follows:

$$\begin{cases} (f\omega)(X) & = f\omega(X) \\ \{(f\omega)(X)\}(p) = f(p)\omega_p(X_p) & [\text{refer to } (3.3)]. \end{cases} \qquad (3.12)$$

Problem 3.2 *Show that* $\omega(fX) = f\omega(X)$, *where* ω *is a* 1-*form on* M, $f \in F(M)$ *and* $X \in \chi(M)$.

Solution: Here

$$\{\omega(fX)\}(p) = \omega_p(fX)(p), \quad \text{by (3.3)}$$
$$= \omega_p f(p)X_p, \quad \text{by (2.26)}$$
$$= f(p)\{\omega(X)\}(p), \quad \text{by (3.3)}$$
$$= \{f\omega(X)\}(p), \quad \text{by (2.26)}$$

Thus $\omega(fX) = f\omega(X), \forall p \in M$.

Problem 3.3 *Let* $X = y\dfrac{\partial}{\partial x} - x\dfrac{\partial}{\partial y} + \dfrac{\partial}{\partial z}$ *be the vector field and* $\omega = zdx + xdz$ *be the* 1-*form on* $\mathbb{R}^3$. *Compute* $\omega(X)$.

Solution: Taking into consideration Remark 3.3 and also (3.6), we obtain

$$\omega(X) = (zdx + xdz)\left(y\frac{\partial}{\partial x} - x\frac{\partial}{\partial y} + \frac{\partial}{\partial z}\right)$$
$$= zy + x.$$

Problem 3.4 *Let* $X = (x^2 + 1)\dfrac{\partial}{\partial x} + (y - 1)\dfrac{\partial}{\partial y}$ *be the vector field and* $\omega = (2xy + y^2 + 1)dx + (x^2 - 1)dy$ *be the* 1-*form on* $\mathbb{R}^2$. *Compute* $\omega(X)$ *at* $(0, 0)$.

Solution: Note that

$$\omega(X) = \{(2xy + y^2 + 1)dx + (x^2 - 1)dy\}\left\{(x^2 + 1)\frac{\partial}{\partial x} + (y - 1)\frac{\partial}{\partial y}\right\}$$

$$= (2xy + y^2 + 1)(x^2 + 1) + (x^2 - 1)(y - 1).$$

Thus, $\omega(X)\big|_{(0,0)} = 2$.

Problem 3.5 Let $X = x\dfrac{\partial}{\partial x} + 2y\dfrac{\partial}{\partial y}$, $Y = xy\dfrac{\partial}{\partial y}$ be the vector field and $\omega = (x + y^2)dx + (x^2 + y)dy$ be the 1-form on $\mathbb{R}^2$. Compute $\omega([X, Y])$.

Solution: Here

$$[X, Y] = \left(x\frac{\partial}{\partial x} + 2y\frac{\partial}{\partial y}\right)\left(xy\frac{\partial}{\partial y}\right) - xy\frac{\partial}{\partial y}\left(x\frac{\partial}{\partial x} + 2y\frac{\partial}{\partial y}\right)$$

$$= xy\frac{\partial}{\partial y}.$$

Thus

$$\omega([X, Y]) = \{(x + y^2)dx + (x^2 + y)dy\}xy\frac{\partial}{\partial y}$$

$$= xy(x^2 + y), \quad \text{by (3.6)}.$$

Problem 3.6 In Problem 3.5, compute $\omega([X, Y])$ at $(1, 1)$.

Solution: Here $\omega([X, Y])\big|_{(1,1)} = 2$.

Problem 3.7 Let $X = -y\dfrac{\partial}{\partial x} - x\dfrac{\partial}{\partial y}$ and $Y = e^x\dfrac{\partial}{\partial x} - y\dfrac{\partial}{\partial y}$ be the two vector fields on $\mathbb{R}^2$. Find a 1-form ω on $\mathbb{R}^2 \setminus \{(0, 0)\}$ such that $\omega(X) = 1$ and $\omega(Y) = 0$.

Solution: Here

$$\det\begin{pmatrix} -y & -x \\ e^x & -y \end{pmatrix} = y^2 + xe^x \neq 0 \text{ on } \mathbb{R}^2 \setminus \{(0, 0)\}.$$

Let $\omega = A(x, y)dx + B(x, y)dy$, where $A, B \in F(\mathbb{R}^2)$ are functions to be determined.

Given that

$$1 = \omega(X) = \{A(x, y)dx + B(x, y)dy\}\left\{-y\frac{\partial}{\partial x} - x\frac{\partial}{\partial y}\right\}$$

$$\therefore \ 1 = \omega(X) = -Ay - Bx \quad \text{and}$$

$$0 = \omega(Y) = Ae^x - By.$$

After a brief calculation, one gets from the last two equations

$$A = -\frac{y}{xe^x + y^2}, \quad B = -\frac{e^x}{xe^x + y^2}.$$

Consequently

$$\omega = -\frac{y}{xe^x + y^2}dx - \frac{e^x}{xe^x + y^2}dy.$$

NOTE: The 1-form ω is on subset U and it is not a 1-form on $\mathbb{R}^2$, as it is not defined at the origin.

Problem 3.8 Let $X = 2\frac{\partial}{\partial x} - \frac{\partial}{\partial y}, Y = e^x\frac{\partial}{\partial y}$ be two vector fields on $\mathbb{R}^2$. Find a 1-form ω on $\mathbb{R}^2$ such that $\omega(X) = 1, \omega(Y) = 0$.

Solution: Note that $\begin{pmatrix} 2 & -1 \\ 0 & e^x \end{pmatrix} = 2e^x \neq 0, \ \forall \ (x, y) \in \mathbb{R}^2$. Let $\omega = A(x, y)dx + B(x, y)dy$, where $A, B \in F(\mathbb{R}^2)$. Now

$$1 = \omega(X) = \{A(x, y)dx + B(x, y)dy\}\left(2\frac{\partial}{\partial x} - \frac{\partial}{\partial y}\right) = 2A - B \text{ and}$$

$$0 = \omega(Y) = -Be^x.$$

Thus $B = 0, A = \frac{1}{2}$. Therefore, $\omega = \frac{1}{2}dx$.

Problem 3.9 Find a 1-form ω on $\mathbb{R}^3$ such that $\omega(X) = 1, \omega(Y) = 0, \omega(Z) = 0$ where $X = xy\frac{\partial}{\partial x} + \frac{\partial}{\partial z}, Y = e^{-x}\frac{\partial}{\partial x} + \frac{\partial}{\partial y}, Z = 2\frac{\partial}{\partial y} + \frac{\partial}{\partial z}$ are vector fields on $\mathbb{R}^3$.

Solution: Note that $\begin{pmatrix} xy & 0 & 1 \\ e^{-x} & 1 & 0 \\ 0 & 2 & 1 \end{pmatrix} = xy + 2e^{-x} \neq 0, \ \forall \ (x, y, z) \in \mathbb{R}^3$. Let

$$\omega = A(x, y, z)dX + B(x, y, z)dy + C(x, y, z)dz, \quad where \quad A, B, C \in F(\mathbb{R}^3).$$

Now

$$1 = \omega(X) \Rightarrow Axy + C = 1$$
$$0 = \omega(Y) \Rightarrow Ae^{-x} + B = 0$$
$$0 = \omega(Z) \Rightarrow 2B + C = 0$$

Solving, we find $A = \dfrac{1}{xy + 2e^{-x}}, B = -\dfrac{e^{-x}}{xy + 2e^{-x}}, C = \dfrac{2e^{-x}}{xy + 2e^{-x}}$. Thus

$$\omega = \frac{dx}{xy + 2e^{-x}} - \frac{e^{-x}dy}{xy + 2e^{-x}} + \frac{2e^{-x}dy}{xy + 2e^{-x}}.$$

Problem 3.10 *Find the subset of $\mathbb{R}^3$ where the vector fields*

$$X = \frac{\partial}{\partial x}, \; Y = \frac{\partial}{\partial x} - \frac{\partial}{\partial y}, \; Z = \frac{\partial}{\partial x} - \frac{\partial}{\partial y} - (1 - x^2)\frac{\partial}{\partial z}$$

are linearly independent. Write the basis $\{\alpha, \beta, \gamma\}$ dual to $\{X, Y, Z\}$ in terms of the basis $\{dx, dy, dz\}$.

Solution: Here

$$\det \begin{pmatrix} 1 & 0 & 0 \\ 1 & -1 & 0 \\ 1 & -1 & -(1 - x^2) \end{pmatrix} = (1 - x^2) \neq 0 \text{ on } \mathbb{R}^3 \setminus \{(x, y, z) | x \neq \pm 1\}.$$

Let us write $\alpha = Adx + Bdy + Cdz$, where A, B, C are functions to be determined. Now from (3.6)

$1 = \alpha(X) = A \Rightarrow A = 1$

$0 = \alpha(Y) = A - B \Rightarrow B = A = 1$

$0 = \alpha(Z) = A - B - C(1 - x^2) \Rightarrow A - B = C(1 - x^2) \Rightarrow C = 0$ as $x \neq \pm 1$.

Thus $\alpha = dx + dy$.

Let us write $\beta = A'dx + B'dy + C'dz$, where A', B', C' are functions to be determined. Now

$$0 = \beta(X) = A' \Rightarrow A' = 0$$

$$1 = \beta(Y) = A' - B' \Rightarrow B' = A' - 1 = -1$$

$$0 = \beta(Z) = A' - B' - C'(1 - x^2) \Rightarrow C' = \frac{1}{1 - x^2}.$$

Thus, $\beta = -dy + \dfrac{1}{1 - x^2}dz$.

Proceeding as above we get, $\gamma = \dfrac{dz}{x^2 - 1}$.

Problem 3.11 *Find the subset of $\mathbb{R}^2$ where the differential forms $\alpha = dx + dy$, $\beta = -dx + (x^2 - 1)dy$ are linearly independent and determine the dual frame $\{X, Y\}$ on it.*

Solution: Here

$$\det \begin{pmatrix} 1 & 1 \\ -1 & x^2 - 1 \end{pmatrix} = x^2 \neq 0 \text{ on } \mathbb{R}^2/\{(x, y)|x \neq 0\}.$$

Let us write

$$X = a\frac{\partial}{\partial x} + b\frac{\partial}{\partial y}, \quad Y = a'\frac{\partial}{\partial x} + b'\frac{\partial}{\partial y}$$

where a, b, a', b' are all functions to be determined. Now

$$1 = X(\alpha) = a + b, \quad a + b = 1$$
$$0 = X(\beta) = -a + b(x^2 - 1), \quad -a + b(x^2 - 1) = 0.$$

From the last two equations, one gets $a = \dfrac{x^2 - 1}{x^2}$. Thus

$$X = \frac{x^2 - 1}{x^2}\frac{\partial}{\partial x} + \frac{1}{x^2}\frac{\partial}{\partial y}.$$

Similarly, one gets $a' = -\dfrac{1}{x^2}, \ b' = \dfrac{1}{x^2}$. Thus

$$Y = -\frac{1}{x^2}\frac{\partial}{\partial x} + \frac{1}{x^2}\frac{\partial}{\partial y}.$$

Problem 3.12 Let f be given in spherical co-ordinate system by $f(r, \theta, \phi) = r \tan \theta$. Consider the point $(r, \theta, \phi) = \left(1, \dfrac{\pi}{4}, 0\right)$. Find the constants A, B, C such that $df\left(1, \dfrac{\pi}{4}, 0\right) = A dr + B d\theta + C d\phi$.

Solution: Now

$$\frac{\partial f}{\partial r} = \tan \theta, \qquad \frac{\partial f}{\partial r}\bigg|_{(1, \frac{\pi}{4}, 0)} = 1$$

$$\frac{\partial f}{\partial \theta} = r \sec^2 \theta, \qquad \frac{\partial f}{\partial \theta}\bigg|_{(1, \frac{\pi}{4}, 0)} = 2$$

$$\frac{\partial f}{\partial \phi} = 0.$$

Thus, $df\left(1, \frac{\pi}{4}, 0\right) = dr + 2d\theta$.

Problem 3.13 *Let us write*

$$e_1 = (1 + y^2)e^z \frac{\partial}{\partial x}, \quad e_2 = (2xy)\frac{\partial}{\partial x} + (1 + y^2)\frac{\partial}{\partial y}, \quad and$$

$$e_3 = -(xy^2)\frac{\partial}{\partial x} - y(1 + y^2)\frac{\partial}{\partial y} - (1 + y^2)\frac{\partial}{\partial z}.$$

(i) *Prove that* $\{e_1, e_2, e_3\}$ *forms a basis of* $\chi(\mathbb{R}^3)$.
(ii) *Find the dual basis* $\{e^1, e^2, e^3\}$ *in terms of* dx, dy, dz.
(iii) *Find* $[e_1, e_2], [e_1, e_3]$.

Solution:

(i) Note that
$$\begin{vmatrix} (1 + y^2)e^z & 0 & 0 \\ 2xy & (1 + y^2) & 0 \\ -xy^2 & -y(1 + y^2) & -(1 + y^2) \end{vmatrix} = -(1 + y^2)^3 e^z \neq 0. \text{ Thus}$$
$\{e_1, e_2, e_3\}$ forms a basis of $\chi(\mathbb{R}^3)$.

(ii) Let us write $e^1 = A dx + B dy + C dz$, where A, B, C are functions to be determined. Now from (3.6), we have

$$1 = e^1(e_1) = A(1 + y^2)e^z$$
$$0 = e^1(e_2) = 2xyA + B(1 + y^2)$$
$$0 = e^1(e_3) = -A(xy^2) - By(1 + y^2) - C(1 + y^2).$$

On solving, one gets

$$A = \frac{1}{(1 + y^2)e^z}, \quad B = -\frac{2xy}{(1 + y^2)e^z}, \quad C = \frac{xy^2}{(1 + y^2)^2 e^z}.$$

Thus
$$e^1 = \frac{1}{(1 + y^2)e^z}dx - \frac{2xy}{(1 + y^2)^2 e^z}dy + \frac{xy^2}{(1 + y^2)^2 e^z}dz.$$

Proceeding as above, we get

$$e^2 = \frac{1}{(1 + y^2)}dy - \frac{y}{(1 + y^2)}dz$$
$$e^3 = -\frac{1}{(1 + y^2)}dz.$$

(iii) It is to be noted that

$$[e_1, e_2] = \left[(1 + y^2)e^z \frac{\partial}{\partial x}, 2xy\frac{\partial}{\partial x} + (1 + y^2)\frac{\partial}{\partial y} \right]$$
$$= \left[(1 + y^2)e^z \frac{\partial}{\partial x}, 2xy\frac{\partial}{\partial x} \right] + \left[(1 + y^2)e^z \frac{\partial}{\partial x}, (1 + y^2)\frac{\partial}{\partial y} \right], \quad \text{by linearity}$$

Taking help of the relation $[fX, gY] = (fg)[X, Y] + \{f(Xg)\}Y - \{g(Yf)\}X$, it can be shown that $[e_1, e_2] = 0$. Similarly, $[e_1, e_3] = (1 + y^2)^2 e^z \dfrac{\partial}{\partial x} = (1 + y^2)e_1$.

Exercises

Exercise 3.2 *If $\omega = zdx + ydz$, compute $\omega(X)$ on $\mathbb{R}^3$ where*

(i) $X = xy\dfrac{\partial}{\partial x} + x^2\dfrac{\partial}{\partial z}$

(ii) $X = y\dfrac{\partial}{\partial y}$

(iii) $X = 2\dfrac{\partial}{\partial x} - \dfrac{\partial}{\partial x} + 3\dfrac{\partial}{\partial x}$

(iv) $X = e^{-x}\dfrac{\partial}{\partial x}$

(v) $X = \dfrac{\partial}{\partial y} + e^x\dfrac{\partial}{\partial z}$

(vi) $X = xy\dfrac{\partial}{\partial y} - \dfrac{x^2}{2}\dfrac{\partial}{\partial x}$.

Exercise 3.3 *Compute $\omega(X)$ at $(1, 0, 1)$ in each cases of Exercise 3.2.*

Exercise 3.4 *Find a 1-form ω on $\mathbb{R}^2 \setminus \{(0, 0)\}$, such that $\omega(X) = 1$ and $\omega(Y) = 0$ where*

(i) $X = xy\dfrac{\partial}{\partial x} + x^2\dfrac{\partial}{\partial y}$, $Y = y\dfrac{\partial}{\partial y}$

(ii) $X = 2\dfrac{\partial}{\partial x} - \dfrac{\partial}{\partial y}$, $Y = e^{-x}\dfrac{\partial}{\partial x}$

(iii) $X = \dfrac{\partial}{\partial y} + e^x\dfrac{\partial}{\partial x}$, $Y = -\dfrac{x^2}{2}\dfrac{\partial}{\partial x} + xy\dfrac{\partial}{\partial y}$.

Exercise 3.5 *Let $\omega = (2xy + y^2 + 1)dx + (x^2 - 1)dy + xdz$ be the 1-form on $\mathbb{R}^3$. Compute $\omega(X)$ and hence $\omega(X)$ at $(1, 0, 0)$ where*

(i) $X = 2xy^2\dfrac{\partial}{\partial x} + (2 + y^2)\dfrac{\partial}{\partial y} + (2 - z)\dfrac{\partial}{\partial z}$.

(ii) $X = 2y^2\dfrac{\partial}{\partial x} + e^z\dfrac{\partial}{\partial y} + (2 - y^2)\dfrac{\partial}{\partial z}$.

(iii) $X = -y(2 + y^2)\dfrac{\partial}{\partial x} - 2z\dfrac{\partial}{\partial y} + (2 + x^2)\dfrac{\partial}{\partial z}$.

Exercise 3.6 *Let $X = x\dfrac{\partial}{\partial x} + 2y\dfrac{\partial}{\partial y}$ and $Y = xy\dfrac{\partial}{\partial x} + y\dfrac{\partial}{\partial y}$ be two vector fields on $\mathbb{R}^2$. Find a 1-form ω on $\mathbb{R}^2 \setminus \{(0, 0)\}$ such that $\omega(X) = 1$ and $\omega(Y) = 0$.*

Exercise 3.7 *Find the basis $\{\alpha, \beta\}$ dual to $\{X, Y\}$ where*

(i) $X = -y\dfrac{\partial}{\partial x} - x\dfrac{\partial}{\partial y}, \quad Y = e^x\dfrac{\partial}{\partial x} - y\dfrac{\partial}{\partial y}.$

(ii) $X = \dfrac{\partial}{\partial x}, \quad Y = \dfrac{\partial}{\partial x} - \dfrac{\partial}{\partial y}.$

(iii) $X = 2y^2\dfrac{\partial}{\partial x} + e^x\dfrac{\partial}{\partial y}, \quad Y = \dfrac{\partial}{\partial y}.$

(iv) $X = e^x\dfrac{\partial}{\partial x} + \dfrac{\partial}{\partial y}, \quad Y = -\dfrac{1}{2}\dfrac{\partial}{\partial x} + y\dfrac{\partial}{\partial y}.$

Exercise 3.8 *Show that the given differential forms*

$$\alpha = \frac{dx}{(2+y^2)e^z} - \frac{2xy}{(2+y^2)^2 e^z}dy,$$

$$\beta = \frac{dy}{(2+y^2)} + \frac{y}{(2+y^2)}dz,$$

$$\gamma = \frac{dz}{(2+y^2)},$$

are linearly independent. Hence find the dual basis $\{X, Y, Z\}$ dual to $\{\alpha, \beta, \gamma\}$ in terms of $\left\{\dfrac{\partial}{\partial x}, \dfrac{\partial}{\partial y}, \dfrac{\partial}{\partial z}\right\}$.

Exercise 3.9 *Find the subset of $\mathbb{R}^2$ where the differential forms*

$$\alpha = \frac{x\,dx}{x^2+y^2} + \frac{y\,dy}{x^2+y^2}, \quad and \quad \beta = -\frac{y\,dx}{x^2+y^2} + \frac{x\,dy}{x^2+y^2}$$

are linearly independent. Hence find the dual basis $\{X, Y\}$ dual to $\{\alpha, \beta\}$ in terms of $\left\{\dfrac{\partial}{\partial x}, \dfrac{\partial}{\partial y}\right\}$.

Exercise 3.10 *Find the subset of $\mathbb{R}^3$ where the differential forms*

$$\alpha = dx + dy, \quad \beta = -dy + \frac{dz}{1-x^2}, \quad \gamma = \frac{dz}{x^2-1}$$

are linearly independent. Hence find the dual basis $\{X, Y, Z\}$ dual to $\{\alpha, \beta, \gamma\}$ in terms of $\left\{\dfrac{\partial}{\partial x}, \dfrac{\partial}{\partial y}, \dfrac{\partial}{\partial z}\right\}$.

Exercise 3.11 *Let $f : \mathbb{R}^3 \to \mathbb{R}$ be given by*

$$f(x, y, z) = (x^2 + y^2 + z^2)^{\frac{3}{2}} + x + 3y.$$

(i) *Write down $df(1, 0, 0)$ in terms of $\{dx, dy, dz\}$.*

(ii) *Express f in spherical co-ordinate (r, θ, ϕ), where*

$$x = r \sin\theta \cos\phi, \quad y = r \sin\theta \sin\phi, \quad z = r \cos\theta.$$

(iii) *Find constants A, B, C such that $df\left(1, \frac{\pi}{2}, 0\right) = A dr + B d\theta + C d\phi$.*

Exercise 3.12 A. *If $\omega = e^x \cos y\, dx + e^x \sin y\, dy$, compute $\omega([X, Y])$ on $\mathbb{R}^2$ where*

(i) $X = x\dfrac{\partial}{\partial x} + y^2\dfrac{\partial}{\partial y}, \ Y = y\dfrac{\partial}{\partial y}$

(ii) $X = xy\dfrac{\partial}{\partial x} + \dfrac{y^2}{2}\dfrac{\partial}{\partial y}, \ Y = 2\dfrac{\partial}{\partial x} - \dfrac{\partial}{\partial y}$

(iii) $X = y\dfrac{\partial}{\partial y}, \ Y = e^x\dfrac{\partial}{\partial x} + \dfrac{\partial}{\partial y}.$

B. *Compute $\omega([X, Y])$ at $(1, 2) \in \mathbb{R}^2$.*

Exercise 3.13 *Find a 1-form ω on $\mathbb{R}^3$ such that $\omega(X) = 1, \omega(Y) = 0, \omega(Z) = 0$ where*

(i) $X = \dfrac{\partial}{\partial x}, Y = \dfrac{\partial}{\partial x} - \dfrac{\partial}{\partial y}, Z = \dfrac{\partial}{\partial x} - \dfrac{\partial}{\partial y} - \dfrac{\partial}{\partial z}$

(ii) $X = (2 + y^2)\dfrac{\partial}{\partial x} + \dfrac{\partial}{\partial y}, Y = \dfrac{\partial}{\partial y} + \dfrac{\partial}{\partial z}, Z = \dfrac{\partial}{\partial x} + \dfrac{\partial}{\partial z}.$

Answers

3.2. (i) $xyz + x^2 y$ (ii) θ (iii) $2z + 3y$ (iv) ze^{-x} (v) ye^x (vi) $-\dfrac{x^2 z}{2}$.

3.3. (i) 0 (ii) 0 (iii) 2 (iv) $\dfrac{1}{e}$ (v) $-\dfrac{1}{2}$.

3.4. (i) $\omega = -\dfrac{1}{xy}\dfrac{\partial}{\partial x}$ (ii) $\omega = -\dfrac{\partial}{\partial y}$ (iii) $\omega = \dfrac{2y}{(x + 2ye^x)}\dfrac{\partial}{\partial x} + \dfrac{x}{(x + 2ye^x)}\dfrac{\partial}{\partial y}.$

3.5. (i) 2 (ii) 2 (iii) 3. **3.6.** $\omega = -\dfrac{dx}{x(2y - 1)} + \dfrac{dy}{2y - 1}$

3.7. (i) $\alpha = -\dfrac{y}{xe^x + y^2}dx - \dfrac{e^x}{xe^x + y^2}dy, \quad \beta = \dfrac{x}{xe^x + y^2}dx - \dfrac{y}{xe^x + y^2}dy.$

(ii) $\alpha = dx + dy, \quad \beta = -dy$ (iii) $\alpha = \dfrac{1}{2y^2}dx, \quad \beta = dy.$

(iv) $\alpha = \dfrac{2y}{1 + 2ye^x}dx + \dfrac{1}{1 + 2ye^x}dy, \quad \beta = -\dfrac{2}{1 + 2ye^x}dx + \dfrac{2e^x}{1 + 2ye^x}dy.$

3.8. $X = (2 + y^2)e^z\dfrac{\partial}{\partial x}, \ Y = 2xy\dfrac{\partial}{\partial x} + (2 + y^2)\dfrac{\partial}{\partial y}, \ Z = -2xy^2\dfrac{\partial}{\partial x} - y(2 + y^2)\dfrac{\partial}{\partial y} + (2 + y^2)\dfrac{\partial}{\partial z}$

3.9. $X = x\dfrac{\partial}{\partial x} + y\dfrac{\partial}{\partial y}, \ Y = -y\dfrac{\partial}{\partial x} + x\dfrac{\partial}{\partial y}$

3.10. $X = \dfrac{\partial}{\partial x}, \quad Y = \dfrac{\partial}{\partial x} - \dfrac{\partial}{\partial y}, \quad Z = \dfrac{\partial}{\partial x} - \dfrac{\partial}{\partial y} + (x^2 - 1)\dfrac{\partial}{\partial z}$

3.11. (i) $4dx + 3dy$ (ii) $r^3 + r \sin\theta \cos\phi + 3r \sin\theta \sin\phi$ (iii) $4dr + 3d\phi$.

3.12. A. (i) $-y^2 e^x \sin y$ (ii) $(x - 2y)e^x \cos y$ (iii) $-e^x \sin y$

   B. (i) $-4e \sin 2$ (ii) $-3e \cos 2$ (iii) $-e \sin 2$.

3.13. (i) $dx + dy$ (ii) $\dfrac{dx}{3 + y^2} + \dfrac{dy}{3 + y^2} - \dfrac{dz}{3 + y^2}$.

3.2 r-form, Exterior Product

An r-form ω is a skew-symmetric mapping

$$\omega : \underbrace{\chi(M) \times \cdots \chi(M)}_{r-times} \to F(M)$$

such that

(i) ω is $\mathbb{R}$-linear

(ii) if σ is a permutation of $1, 2, 3, \ldots, r$ with $(1, 2, 3, \ldots, r) \to (\sigma(1), \sigma(2), \ldots, \sigma(r))$, then

$$\omega(X_1, X_2, \ldots, X_r) = \frac{1}{r!} \sum_\sigma (sgn\ \sigma)\ \omega(X_{\sigma(1)}, X_{\sigma(2)}, \ldots, X_{\sigma(r)}) \qquad (3.13)$$

where $sgn\ \sigma$(pronounced as **signum** σ) is $+1$ or -1 according as σ is even or odd permutation.

The product of two skew-symmetric form is called the **exterior product** or **Grassmann product**, as introduced by H.G Grassmann or **Wedge Product**, as a wedge '$\wedge$' is used to denote this product. We are going to give the formal definition.

Remark 3.5 By convention, a zero-form is a function.

If ω is a r-form and μ is a s-form, then the exterior product or wedge product of ω and μ, denoted by $\omega \wedge \mu$ is a $(r + s)$-form defined as

$$(\omega \wedge \mu)(X_1, X_2, \ldots, X_r, X_{r+1}, \ldots, X_{r+s}) \qquad (3.14)$$

$$= \frac{1}{(r + s)!} \sum_\sigma (sgn\ \sigma)\ \omega(X_{\sigma(1)}, X_{\sigma(2)}, \ldots, X_{\sigma(r)}) \mu(X_{\sigma(r+1)}, X_{\sigma(r+2)}, \ldots, X_{\sigma(r+s)})$$

where σ ranges over the permutation $(1, 2, 3, \ldots, r + s)$, $X_i \in \chi(M)$, $i = 1, 2, 3, \ldots, r + s$.

For convenience, we write

$$f \wedge g = fg; \ \forall\ f, g \in F(M). \qquad (3.15)$$

It can be shown that, for a r-form ω

$$\begin{cases} (f \wedge \omega)(X_1, X_2, \ldots, X_r) = f\omega(X_1, X_2, \ldots, X_r) \\ (\omega \wedge f)(X_1, X_2, \ldots, X_r) = f\omega(X_1, X_2, \ldots, X_r) \end{cases} \tag{3.16}$$

Again, if ω and μ are 1-forms, then

$$(\omega \wedge \mu)(X_1, X_2) = \frac{1}{2}\{\omega(X_1)\mu(X_2) - \omega(X_2)\mu(X_1)\}. \tag{3.17}$$

The exterior product obeys the following properties:

$$\begin{cases} \omega \wedge \omega = 0, \ \omega \wedge \mu = (-1)^{rs}\mu \wedge \omega, \ \mu \text{ being } s\text{-form} \\ f\omega \wedge \mu = f(\omega \wedge \mu) = \omega \wedge f\mu \\ f\omega \wedge g\mu = fg\,\omega \wedge \mu \\ (\omega + \mu) \wedge \gamma = \omega \wedge \gamma + \mu \wedge \gamma, \quad \omega \wedge (\mu + \gamma) = \omega \wedge \mu + \omega \wedge \gamma. \end{cases} \tag{3.18}$$

Problem 3.14 *Given 1-form $\omega = fdx - gdy + hdz$ and $\mu = f'dx + g'dy$ in $\mathbb{R}^3$. Compute $\omega \wedge \omega$ and show that $\omega \wedge \mu = -\mu \wedge \omega$. Prove that each $\omega \wedge \mu$, $\mu \wedge \omega$ is a 2-form.*

Solution: Taking help of (3.18), the result follows immediately.

Problem 3.15 *Let V be a vector space of dimension 3 over $\mathbb{R}$. Let $\{e_1, e_2, e_3\}$ be a basis of V and $\{e^1, e^2, e^3\}$ be its dual basis. Let α, β be co-vectors, so that $\alpha = a_1e^1 + a_2e^2 + a_3e^3$, $\beta = b_1e^1 + b_2e^2 + b_3e^3$ where each a_i, $b_i \in F(V)$. Show that the components of $\alpha \wedge \beta = \alpha \times \beta$.*

Solution: Here

$$\begin{aligned} \alpha \wedge \beta &= (a_1e^1 + a_2e^2 + a_3e^3) \wedge (b_1e^1 + b_2e^2 + b_3e^3) \\ &= (a_1b_2 - a_2b_1)e^1 \wedge e^2 + (a_2b_3 - a_3b_2)e^2 \wedge e^3 + (a_3b_1 - a_1b_3)e^3 \wedge e^1. \end{aligned}$$

Thus the components of $\alpha \wedge \beta = (a_1b_2 - a_2b_1, a_2b_3 - a_3b_2, a_3b_1 - a_1b_3)$. Again

$$\alpha \times \beta = (a_1b_2 - a_2b_1, a_2b_3 - a_3b_2, a_3b_1 - a_1b_3).$$

Hence the proof.

Problem 3.16 *Let a 1-form α and 2-form β be given on $\mathbb{R}^3$ as $\alpha = adx + bdy + cdz$, $\beta = a' \, dx \wedge dy + b' \, dy \wedge dz + c' \, dz \wedge dx$. Compute $\alpha \wedge \beta$.*

Solution: Note that

$$\alpha \wedge \beta = (adx + bdy + cdz) \wedge (a'\, dx \wedge dy + b'\, dy \wedge dz + c'\, dz \wedge dx)$$
$$= ab'\, dx \wedge dy \wedge dz + bc'\, dy \wedge dz \wedge dx + ca'\, dz \wedge dx \wedge dy.$$

As $dx \wedge dz \wedge dx = -dx \wedge dx \wedge dz = 0$ and so on, we have

$$\alpha \wedge \beta = (ab' + bc' + ca')dx \wedge dy \wedge dz,$$

which is a 3-form.

Theorem 3.2 *In terms of a local co-ordinate system* $(x^1, x^2, \ldots, x^n)$ *in a neighbourhood* U *of* p *of* M, *an* r-*form can be expressed uniquely as*

$$\omega = \sum_{i_1, i_2, \ldots, i_r} f_{i_1 i_2 \cdots i_r} dx^{i_1} \wedge dx^{i_2} \wedge \cdots \wedge dx^{i_r}, \quad i_1 < i_2 < \ldots < i_r \qquad (3.19)$$

where $f_{i_1 i_2 \cdots i_r}$ *are differentiable functions.*

Proof As the set $\{dx^{i_1} \wedge dx^{i_2} \wedge \cdots \wedge dx^{i_r}\}$ is a basis of $\mathfrak{D}_r(M^n)$, from (3.14) we find

$$(dx^{i_1} \wedge dx^{i_2} \wedge \cdots \wedge dx^{i_r})(X_1, X_2, \ldots, X_r) = \frac{1}{r!} \sum_\sigma (sgn\,\sigma) dx^{i_1}(X_{\sigma(1)}) \ldots dx^{i_r}(X_{\sigma(r)}),$$

where $i_1 < i_2 < \ldots < i_r$.

Let us write,

$$X_i = \sum_{i_m} \xi_i^{j_m} \frac{\partial}{\partial x^{j_m}}, \quad i = 1, 2, 3, 4, \ldots, r,$$

where each $\xi_i^{i_m}$ is C^∞ function. Using this in the foregoing equation, we obtain

$$(dx^{i_1} \wedge dx^{i_2} \wedge \cdots \wedge dx^{i_r})(X_1, X_2, \ldots, X_r)$$

$$= \frac{1}{r!} \sum_{(\sigma)} (sgn\,\sigma) dx^{i_1} \left(\sum \xi_{\sigma(1)}^{j_m} \frac{\partial}{\partial x^{j_m}} \right) \cdots dx^{i_r} \left(\sum \xi_{\sigma(r)}^{j_m} \frac{\partial}{\partial x^{j_m}} \right)$$

$$= \frac{1}{r!} \sum_{(\sigma)} (sgn\,\sigma) \xi_{\sigma(1)}^{i_1} \cdots \xi_{\sigma(r)}^{i_r}, \quad i_1 < i_2 < \ldots < i_r \text{ (by (3.6))}.$$

From (3.13), we see that

$$\omega(X_1, X_2, \ldots, X_r) = \frac{1}{r!} \sum_{(\sigma)} (sgn \ \sigma) \ \omega \left(\sum \xi^{jm}_{\sigma(1)} \frac{\partial}{\partial x^{jm}}, \ldots, \sum \xi^{jm}_{\sigma(r)} \frac{\partial}{\partial x^{jm}} \right)$$

$$= \frac{1}{r!} \sum_{(\sigma)} (sgn \ \sigma) \sum_{i_1, i_2, \ldots, i_r} \xi^{i_1}_{\sigma(1)} \cdots \xi^{i_r}_{\sigma(r)} \omega \left(\frac{\partial}{\partial x^{i_1}}, \ldots, \frac{\partial}{\partial x^{i_r}} \right)$$

$$= \frac{1}{r!} \sum_{(\sigma)} (sgn \ \sigma) \sum_{i_1, i_2, \ldots, i_r} \xi^{i_1}_{\sigma(1)} \cdots \xi^{i_r}_{\sigma(r)} f_{i_1 i_2 \cdots i_r}, \ \text{say, as defined in (3.7)}$$

$$= \sum_{\substack{i_1, i_2, \ldots, i_r \\ i_1 < i_2 < \ldots < i_r}} (dx^{i_1} \wedge dx^{i_2} \wedge \cdots \wedge dx^{i_r})(X_1, X_2, \ldots, X_r) f_{i_1 i_2 \cdots i_r}, \ \text{from above.}$$

Thus $\quad \omega = \displaystyle\sum_{\substack{i_1, i_2, \ldots, i_r \\ i_1 < i_2 < \ldots < i_r}} f_{i_1 i_2 \cdots i_r} \ dx^{i_1} \wedge dx^{i_2} \wedge \cdots \wedge dx^{i_r}, \quad \text{for each} \quad X_i, \ i = 1, 2,$

$3, \ldots, r$.

Theorem 3.3 *In terms of a local co-ordinate system* $(x^1, x^2, \ldots, x^n)$ *in a neighbourhood* U *of* p *of a manifold, let* $f^i, i = 1, 2, 3, \ldots, r$, *be smooth functions on* U. *Then*

$$df^1 \wedge df^2 \wedge \cdots \wedge df^r = \sum_{\substack{i_1, i_2, \ldots, i_r \\ i_1 < i_2 < \ldots < i_r}} \frac{\partial(f^1, f^2, \ldots, f^r)}{\partial(x^{i_1}, x^{i_2}, \ldots, x^{i_r})} dx^{i_1} \wedge dx^{i_2} \wedge \cdots \wedge dx^{i_r}.$$

Proof Each $df^i, i = 1, 2, 3, \ldots, r$, is a 1-form and $df^1 \wedge df^2 \wedge \cdots \wedge df^r$ is a r-form and in view of Theorem 3.2, we write

$$df^1 \wedge df^2 \wedge \cdots \wedge df^r = \sum_{j_1, j_2, \ldots, j_r} F_{j_1 j_2 \cdots j_r} dx^{j_1} \wedge dx^{j_2} \wedge \cdots \wedge dx^{j_r}, \quad j_1 < j_2 < \ldots < j_r,$$

where $F_{j_1 j_2 \cdots j_r}$ are differentiable functions to be determined. In view of (3.10), we find

$$df = \sum_j \frac{\partial f}{\partial x^j} dx^j \Rightarrow df^i = \sum_k \frac{\partial f^i}{\partial x^k} dx^k.$$

Now

$$df^i \left(\frac{\partial}{\partial x^j} \right) = \sum_k \frac{\partial f^i}{\partial x^k} dx^k \left(\frac{\partial}{\partial x^j} \right) = \sum_k \frac{\partial f^i}{\partial x^k} \delta^k_j = \frac{\partial f^i}{\partial x^j}.$$

Hence from the left hand side of the above expression, we have

$$(df^1 \wedge df^2 \wedge \cdots \wedge df^r) \left(\frac{\partial}{\partial x^{i_1}}, \frac{\partial}{\partial x^{i_2}}, \ldots, \frac{\partial}{\partial x^{i_r}} \right) = \frac{\partial(f^1, f^2, \ldots, f^r)}{\partial(x^{i_1}, x^{i_2}, \ldots, x^{i_r})}.$$

Furthermore, from the right hand side, we get

$$\sum_{j_1, j_2, \ldots, j_r} F_{j_1 j_2 \cdots j_r} dx^{j_1} \wedge dx^{j_2} \wedge \cdots \wedge dx^{j_r} \left(\frac{\partial}{\partial x^{i_1}}, \frac{\partial}{\partial x^{i_2}}, \ldots, \frac{\partial}{\partial x^{i_r}} \right) = \sum_{j_1, j_2, \ldots, j_r} F_{j_1 j_2 \cdots j_r} \delta_{j_1 \cdots j_r}^{i_1 \cdots i_r}$$

$$= F_{i_1 i_2 \cdots i_r}.$$

Thus

$$\frac{\partial(f^1, f^2, \ldots, f^r)}{\partial(x^{i_1}, x^{i_2}, \ldots, x^{i_r})} = F_{i_1 i_2 \cdots i_r}.$$

Consequently,

$$df^1 \wedge df^2 \wedge \cdots \wedge df^r = \sum_{\substack{j_1, j_2, \ldots, j_r \\ j_1 < j_2 < \ldots < j_r}} \frac{\partial(f^1, f^2, \ldots, f^r)}{\partial(x^{j_1}, x^{j_2}, \ldots, x^{j_r})} dx^{j_1} \wedge dx^{j_2} \wedge \cdots \wedge dx^{j_r}$$

i.e. $df^1 \wedge df^2 \wedge \cdots \wedge df^r = \sum_{\substack{i_1, i_2, \ldots, i_r \\ i_1 < i_2 < \ldots < i_r}} \frac{\partial(f^1, f^2, \ldots, f^r)}{\partial(x^{i_1}, x^{i_2}, \ldots, x^{i_r})} dx^{i_1} \wedge dx^{i_2} \wedge \cdots \wedge dx^{i_r}.$

This completes the proof.

Proposition 3.1 *If ω and μ are C^∞-forms on M, then $\omega \wedge \mu$ is also C^∞.*

Proof From (3.8) and Theorem 3.1, we know that

$$\omega = \sum_i f_i dx^i, \quad \mu = \sum_j g_j dx^j,$$

where $f_i, g_j \in F(M)$ for each i, j. By Remark 3.2, we say that ω, μ are C^∞-forms on M, if each f_i, g_j are C^∞-functions on M. Now

$$\omega \wedge \mu = \sum_{i,j} f_i g_j dx^i \wedge dx^j, \quad \text{Theorem 3.3.}$$

Since $\sum_{i,j} f_i g_j$ are C^∞-functions on M, by Remark 3.2, we can claim that $\omega \wedge \mu$ is also C^∞-forms on M.

Exercises

Exercise 3.14 *If ω is a 1-form and μ is a 2 form, show that*

$$(\omega \wedge \mu)(X_1, X_2, X_3) = \frac{1}{3}\{\omega(X_1)\mu(X_2, X_3) + \omega(X_2)\mu(X_3, X_1) + \omega(X_3)\mu(X_1, X_2)\}.$$

Exercise 3.15 *Compute $\omega \wedge \mu$ where*

(i) $\omega = xdx - ydy, \quad \mu = ydx + hdy$
(ii) $\omega = xdx + ydy, \quad \mu = gdx$

(iii) $\omega = zdx + xdz, \quad \mu = xdx + ydy + zdz$

(iv) $\omega = x\, dy \wedge dz, \quad \mu = gdx + hdz.$

Exercise 3.16 *Compute*

(i) $(2du^1 + du^2) \wedge (du^1 - du^2).$

(ii) $(6du^1 \wedge du^2 + 27du^1 \wedge du^3) \wedge (du^1 + du^2 + du^3).$

(iii) $\theta \wedge \phi \wedge \psi$ where $\theta = zdy, \quad \phi = xdx + ydy, \quad \psi = zdx - ydz.$

Exercise 3.17 *If* $\alpha = -du^1 + du^2 - 2du^3$ *is a 1-form and* β *is a 2-form given by* $\beta = 2\, du^1 \wedge du^3 - du^2 \wedge du^3$, *compute* $\alpha \wedge \beta \wedge \alpha.$

Answers

    3.14. Use (3.14).

    3.15. (i) $(xh + y^2)\, dx \wedge dy.$ (ii) $-yg\, dx \wedge dy$

          (iii) $(z^2 - x^2)\, dx \wedge dz + zy\, dx \wedge dy - xy\, dx \wedge dy$ (iv) $xg\, dx \wedge dy \wedge dz.$

    3.16. (i) $-3\, du^1 \wedge du^2.$ (ii) $-21\, du^1 \wedge du^2 \wedge du^3.$ (iii) $xyz\, dx \wedge dy \wedge dz.$

    3.17. 0 (Zero).

Remark 3.6 Let $\mathfrak{D}_r(M^n)$ denote the collection of all r-forms in M^n. By virtue of Theorem 3.1, the set $\{dx^{i_1} \wedge dx^{i_2} \wedge \cdots \wedge dx^{i_r} : 1 \leq i_1 < i_2 < \cdots < i_r \leq n\}$ forms a basis of $\mathfrak{D}_r(M^n)$.

Remark 3.7 The collection of all differential forms with respect to wedge product, forms an algebra, called the EXTERIOR ALGEBRA. Now a days it is also termed as GRASSMANN ALGEBRA, as R.G. Grassmann developed this powerful concept.

Remark 3.8 For a manifold M^4, every point $p \in M$, has coordinates of the form $p = (x^1, x^2, x^3, x^4)$. Thus the basis set of $\mathfrak{D}_1(M^4)$ contains $^4C_1$ elements viz $\{dx^1, dx^2, dx^3, dx^4\}$;

that of $\mathfrak{D}_2(M^4)$ is $^4C_2$ i.e. $\{dx^1 \wedge dx^2, \ dx^1 \wedge dx^3, \ dx^1 \wedge dx^4, \ dx^2 \wedge dx^3, \ dx^2 \wedge dx^4, \ dx^3 \wedge dx^4\}$;

that of $\mathfrak{D}_3(M^4)$ is $^4C_3$ i.e. $\{dx^1 \wedge dx^2 \wedge dx^3, \ dx^1 \wedge dx^2 \wedge dx^4, \ dx^1 \wedge dx^3 \wedge dx^4, \ dx^2 \wedge dx^3 \wedge dx^4\}$;

and that of $\mathfrak{D}_4(M^4)$ is $^4C_4$ given by $\{dx^1 \wedge dx^2 \wedge dx^3 \wedge dx^4\}$. **However,** $\mathfrak{D}_5(M^4)$ **does not exist.** Hence any,

$$\omega \in \mathfrak{D}_1(M^4) \text{ is of the form } \omega = \sum_{i=1}^{4} f_i dx^i;$$

$$\omega \in \mathfrak{D}_2(M^4) \text{ is of the form } \omega = \sum_{ij} f_{ij} dx^i \wedge dx^j, \ 1 \leq i < j \leq 4;$$

$$\omega \in \mathfrak{D}_3(M^4) \text{ is of the form } \omega = \sum_{i,j,k} f_{ijk} dx^i \wedge dx^j \wedge dx^k, \ 1 \leq i < j < k \leq 4;$$

$$\omega \in \mathfrak{D}_4(M^4) \text{ is of the form } \omega = f_{1234} dx^1 \wedge dx^2 \wedge dx^3 \wedge dx^4.$$

In Sect. 3.1, the transition formula for a 1-form has been given by the (3.11). Now we will focus to find the **Transition Formula for a 2-We write it asform**.

Let $(U; x^1, x^2, \ldots, x^n)$ and $(V; y^1, y^2, \ldots, y^n)$ be two charts on a manifold M such that $U \cap V \neq \phi$. If ω is a 2-form, then by (3.19), we have

$$\omega = \sum_{\substack{i,j \\ i<j}} f_{ij} dx^i \wedge dx^j = \sum_{\substack{m,n \\ m<n}} g_{mn} dy^m \wedge dy^n, \text{ where each } f_{ij}, g_{mn} \in F(M)$$

$$= \sum_{\substack{i,j \\ i<j}} \sum_{\substack{m,n \\ m<n}} g_{mn} \frac{\partial(y^m, y^n)}{\partial(x^i, x^j)} dx^i \wedge dx^j, \text{ by Theorem 3.3}$$

$$f_{ij} = \sum_{\substack{m,n \\ m<n}} \frac{\partial(y^m, y^n)}{\partial(x^i, x^j)} g_{mn}, \quad i, j = 1, 2, 3, \ldots, n, \tag{3.20}$$

as $\{dx^i \wedge dx^j : i, j = 1, 2, 3, \ldots, n; 1 \leq i < j \leq n\}$ is a basis of 2-form.

Problem 3.17 *Let $\{\omega_1, \omega_2, \omega_3\}$ be a set of linearly independent 1-form on a smooth manifold M. Define 1-form $\mu_i, i = 1, 2, 3$ as $\mu_i = \sum_{j=1}^{3} a_{ij} \omega_j, i = 1, 2, 3$. Show that*

$$\mu_1 \wedge \mu_2 \wedge \mu_3 = \det(a_{ij}) \omega_1 \wedge \omega_2 \wedge \omega_3.$$

Solution: We write

$$\mu_1 = \sum a_{1j} \omega_j, \quad \mu_2 = \sum a_{2j} \omega_j \quad \mu_3 = \sum a_{3j} \omega_j.$$

Thus

$$\mu_1 = a_{11} \omega_1 + a_{12} \omega_2 + a_{13} \omega_3$$
$$\mu_2 = a_{21} \omega_1 + a_{22} \omega_2 + a_{23} \omega_3$$
$$\mu_3 = a_{31} \omega_1 + a_{32} \omega_2 + a_{33} \omega_3.$$

Now

$$\mu_1 \wedge \mu_2 = (a_{11}a_{22} - a_{12}a_{21})\omega_1 \wedge \omega_2 + (a_{11}a_{23} - a_{13}a_{21})\omega_1 \wedge \omega_3$$
$$+ (a_{12}a_{23} - a_{13}a_{22})\omega_2 \wedge \omega_3,$$

as $\omega_1 \wedge \omega_1 = 0$, $\omega_i \wedge \omega_j = -\omega_j \wedge \omega_i$. Again Now

$$\mu_1 \wedge \mu_2 \wedge \mu_3 = a_{33}(a_{11}a_{22} - a_{12}a_{21})\omega_1 \wedge \omega_2 \wedge \omega_3$$
$$+ a_{32}(a_{11}a_{23} - a_{13}a_{21})\omega_1 \wedge \omega_2 \wedge \omega_3$$
$$+ a_{31}(a_{12}a_{23} - a_{13}a_{22})\omega_1 \wedge \omega_2 \wedge \omega_3$$
$$= \det(a_{ij})\omega_1 \wedge \omega_2 \wedge \omega_3, \quad \text{as } \omega_i \wedge \omega_j = -\omega_j \wedge \omega_i, i, j = 1, 2, 3.$$

Problem 3.18 *(Cartan Lemma) Let $k < n$ and $\{\omega_1, \omega_2, \ldots, \omega_k\}$ be 1-forms on M^n which are linearly independent pointwise. Let μ_i be k number of 1-forms on M satisfying*

$$\sum_{i=1}^{k} \mu_i \wedge \omega_i = 0.$$

Prove that there exists C^∞ functions A_{ij} on M^n such that

$$\mu_i = \sum_{j=1}^{k} A_{ij}\omega_j \quad \text{with} \quad A_{ij} = A_{ji}, \quad i = 1, 2, 3, 4, \ldots, k.$$

Solution: As $\{\omega_1, \omega_2, \ldots, \omega_k\}$ is k-independent 1-forms on M, we complete the basis of $\mathfrak{D}_1(M)$ by taking 1-forms $\omega_{k+1}, \ldots, \omega_n$. Consequently, any 1-form $\mu_i, i = 1, 2, \ldots, k$ can be expressed as

$$\mu_i = \sum_{m=1}^{k} A_{im}\omega_m + \sum_{p=k+1}^{n} B_{ip}\omega_p, \quad i = 1, 2, 3, \ldots, k,$$

where each A_{im}, B_{ip} is C^∞-function. Given that $\sum_{i=1}^{k} \mu_i \wedge \mu_i = 0$ i.e.

$$\left(\sum_m A_{1m}\omega_m + \sum_p B_{1p}\omega_p\right) \wedge \omega_1 + \cdots + \left(\sum_m A_{km}\omega_m + \sum_p B_{kp}\omega_p\right) \wedge \omega_k = 0.$$

Using the properties

$$\omega_i \wedge \omega_i = 0 \quad \text{and} \quad \omega_i \wedge \omega_j = -\omega_j \wedge \omega_i,$$

one gets

$$\sum_{\substack{i,j \\ i<j\leq k}} (A_{ij} - A_{ji})\omega_i \wedge \omega_j + \sum_{\substack{i\leq k \\ j>k}} B_{ij}\omega_i \wedge \omega_j = 0.$$

As $\{\omega_1, \omega_2, \ldots, \omega_n\}$ is a basis of $\mathfrak{D}_1(M)$, we must have $A_{ij} - A_{ji} = 0$ and $B_{ij} = 0$. Consequently, $\mu_i, i = 1, 2, \ldots, k$ can be expressed as

$$\mu_i = \sum_{j=1}^{k} A_{im}\omega_j, \quad \text{with} \quad A_{ij} = A_{ji}.$$

Exercise

Exercise 3.18 *Show that a set of 1-forms $\{\omega_1, \omega_2, \ldots, \omega_k\}$ is linearly dependent if and only if $\omega_1 \wedge \omega_2 \wedge \cdots \wedge \omega_k = 0$.*

Exercise 3.19 *Let $\{\omega_1, \ldots, \omega_n\}$ be a set of linearly independent 1-forms on a smooth manifold. Let $\mu_i, i = 1, 2, 3, \ldots, n$ be 1-forms on M satisfying*

$$\mu_i = \sum_{j=1}^{n} a_{ij} \omega_j.$$

Show that $\mu_1 \wedge \mu_2 \wedge \mu_3 \wedge \ldots \wedge \mu_n = \det(a_{ij}) \, \omega_1 \wedge \omega_2 \wedge \omega_3 \wedge \ldots \wedge \omega_n$.

3.3   Exterior Differentiation

As discussed in Remark 3.7 of the last section, we write $\mathfrak{D} = \bigoplus_{r=0}^{n} \mathfrak{D}_r(M^n)$, as the **Exterior Algebra**, with respect to the wedge product in M^n. It is interesting to note that

$$\dim \mathfrak{D} =^n C_0 +^n C_1 +^n C_2 + \ldots +^n C_n$$
$$= (1+1)^n, \text{ by Binomial Theorem}$$
$$= 2^n.$$

We are now going to define the exterior derivative which is a linear mapping, denoted by d, on $\mathfrak{D}$ as follows:

$$
\begin{cases}
(i) \ d(\mathfrak{D}_r) \subset \mathfrak{D}_{r+1}; \\
(ii) \ \text{for } f \in \mathfrak{D}_0, \ df \text{ is the total differential;} \\
(iii) \ \text{if } \omega \in \mathfrak{D}_r, \mu \in \mathfrak{D}_s, \text{ then } d(\omega \wedge \mu) = d\omega \wedge \mu + (-1)^r \omega \wedge d\mu; \\
(iv) \ d^2 = 0.
\end{cases}
\qquad (3.21)
$$

Hence from (3.19), we see that

$$d\omega = \sum_{i_1, i_2, \ldots, i_r} df_{i_1 i_2 \ldots i_r} \wedge dx^{i_1} \wedge \cdots \wedge dx^{i_r}, \ i_1 < i_2 < \ldots < i_r. \qquad (3.22)$$

Problem 3.19 *Find the exterior derivative of the following:*

(i) $f = x^2 y^3 z$ **(ii)** $f g$.

Solution:

(i) Here $df = 2xy^3z\,dx + 6x^2y^2z\,dy + x^2y^3\,dz$. Thus df is a 1-form, verifying (3.21)(ii)

(ii) We know $df(X) = Xf$. Thus

$$d(fg)(X) = X(fg)$$
$$= g(Xf) + f(Xg)$$
$$= g\,df(X) + f\,dg(X),$$

i.e. $d(fg) = g\,df + f\,dg$.

Problem 3.20 *Verify (3.21)(iv) for $\omega = x^2y\,dx$.*

Solution: Note that

$$d\omega = d(x^2y\,dx)$$
$$= d(x^2y) \wedge dx, \text{ see (3.22)}$$
$$= (2xy\,dx + x^2\,dy) \wedge dx$$
$$= x^2\,dy \wedge dx, \text{ as } dx \wedge dx = 0.$$

Again

$$d(d\omega) = d(x^2\,dy \wedge dx)$$
$$= d(x^2) \wedge dy \wedge dx, \text{ see (3.22)}$$
$$= 2x\,dx \wedge dy \wedge dx$$
$$= -2x\,dx \wedge dx \wedge dy$$
$$= 0.$$

Problem 3.21 *Find the exterior derivative of:*

(i) $\omega_1 = 2xdx + (x + y)dy$
(ii) $\omega_2 = x^2ydx - xz^3dy + 3xydz$
(iii) $\omega_3 = (x^2 - y^2)dx \wedge dy$
(iv) $\omega_4 = x^2yz\,dy \wedge dz - 2xyz\,dz \wedge dx + xyz^3\,dx \wedge dy.$

Solution:

(i) Here $\omega_1 = 2xdx + (x + y)dy$. Therefore

$$d\omega_1 = d(2xdx) + d\{(x + y)dy\}, \text{ by linearity}$$
$$= d(2x) \wedge dx + (-1)^0 2x\,d(dx) + d(x + y) \wedge dy + (-1)^0(x + y)d(dy), \text{ by (iii) of (3.21)},$$

as the function is assumed to be a 0 − form.

$$= (2dx) \wedge dx + (dx + dy) \wedge dy, \text{ by (iv) of (3.21)}$$
$$= 0 + dx \wedge dy + 0$$
$$= dx \wedge dy.$$

(ii) Note that $\omega_2 = x^2y\,dx - xz^3\,dy + 3xy\,dz$. Hence

$$
\begin{aligned}
d\omega_2 &= d(x^2y) \wedge dx - d(xz^3) \wedge dy + d(3xy) \wedge dz, \text{ by (3.21)} \\
&= (2xy\,dx + x^2\,dy) \wedge dx - (z^3\,dx + 3xz^2\,dz) \wedge dy + (3y\,dx + 3x\,dy) \wedge dz \\
&= x^2\,dy \wedge dx - z^3\,dx \wedge dy - 3xz^2\,dz \wedge dz + 3y\,dx \wedge dz + 3x\,dy \wedge dz \\
&= (-x^2 - z^3)dx \wedge dy + (3xz^2 + 3x)dy \wedge dz - 3y\,dz \wedge dx.
\end{aligned}
$$

(iii) Here $\omega_3 = (x^2 - y^2)dx \wedge dy$. Therefore

$$
\begin{aligned}
d\omega_3 &= d(x^2 - y^2) \wedge dx \wedge dy, \text{ by (3.21)} \\
&= (2x\,dx - 2y\,dy) \wedge dx \wedge dy \\
&= 2x\,dx \wedge dx \wedge dy - 2y\,dy \wedge dx \wedge dy \\
&= 0 + 2y\,dy \wedge dy \wedge dx \\
&= 0.
\end{aligned}
$$

(iv) Similarly

$$
d\omega_4 = (2xyz - 2xz + 3xyz^2)\,dx \wedge dy \wedge dz.
$$

Problem 3.22 *Let ω be a 1-form on a manifold M. Consider a nowhere vanishing function $f : M \to \mathbb{R}$ such that $d(f\omega) = 0$. Prove that $\omega \wedge d\omega = 0$.*

Solution: From the definition,

$$
\begin{aligned}
d(f\omega) &= df \wedge \omega + f \wedge d\omega, \text{ by (3.21)} \\
\therefore \quad 0 &= df \wedge \omega + + f\,d\omega, \text{ by (3.16)} \\
\text{or } d\omega &= -\frac{1}{f}(df \wedge \omega).
\end{aligned}
$$

$$
\begin{aligned}
\text{Now } \omega \wedge d\omega &= \omega \wedge -\frac{1}{f}(df \wedge \omega) \\
&= \frac{1}{f}\,\omega \wedge (\omega \wedge df), \text{ by (3.18)} \\
&= 0, \text{ by (3.18)}.
\end{aligned}
$$

Problem 3.23 *For any 1-form ω, is $\omega \wedge d\omega = 0$? Justify your answer with an example.*

Solution: Let $\omega \in \mathfrak{D}_1(\mathbb{R}^3)$ be such that

$$
\omega = f\,dx^1 + g\,dx^2 + h\,dx^3, \text{ where } f, g, h \in F(\mathbb{R}^3).
$$

Then
$$d\omega = df \wedge dx^1 + dg \wedge dx^2 + dh \wedge dx^3, \text{ by (3.21)}.$$

Thus
$$\omega \wedge d\omega = f dx^1 \wedge dg \wedge dx^2 + f dx^1 \wedge dh \wedge dx^3 + g dx^2 \wedge df \wedge dx^1$$
$$+ g dx^2 \wedge dh \wedge dx^3 + h dx^3 \wedge df \wedge dx^1 + h dx^3 \wedge dg \wedge dx^2$$
$$\neq 0, \text{ always}.$$

Problem 3.24 *Consider the 1-forms* $\omega_1, \omega_2, \omega_3$ *defined by*

$$\omega_1 = h dx^1 - x^1 dh - x^2 dx^3 + x^3 dx^2$$
$$\omega_2 = h dx^2 - x^2 dh - x^3 dx^1 + x^1 dx^3$$
$$\omega_3 = h dx^3 - x^3 dh - x^1 dx^2 + x^2 dx^1$$

in   terms   of   a   local   co-ordinate   system (x^1, x^2, x^3) where $h = \sqrt{1 - (x^1)^2 - (x^2)^2 - (x^3)^2}$. Show that $d\omega_1 = -2\omega_2 \wedge \omega_3$.

Solution: Note that

$$d\omega_1 = -2 dx^1 \wedge dh - 2 dx^2 \wedge dx^3,$$
$$dh = -\frac{x^1 dx^1 + x^2 dx^2 + x^3 dx^3}{\sqrt{1 - (x^1)^2 - (x^2)^2 - (x^3)^2}}.$$

Thus
$$dx^1 \wedge dh = -\frac{x^2 dx^1 \wedge dx^2 + x^3 dx^1 \wedge dx^3}{\sqrt{1 - (x^1)^2 - (x^2)^2 - (x^3)^2}}.$$

Again
$$\omega_2 \wedge \omega_3 = h^2 dx^2 \wedge dx^3 - x^3 h dx^2 \wedge dh + x^2 h dx^2 \wedge dx^1 - x^2 h dh \wedge dx^3$$
$$+ x^1 x^2 dh \wedge dx^2 - (x^2)^2 dh \wedge dx^1 - x^3 h dx^1 \wedge dx^3$$
$$+ (x^3)^2 dx^1 \wedge dh + x^1 x^3 dx^1 \wedge dx^2 - x^1 x^3 dx^3 \wedge dh$$
$$- (x^1)^2 dx^3 \wedge dx^2 + x^1 x^2 dx^3 \wedge dx^1.$$

It is to be noted that

$$h^2\, dx^2 \wedge dx^3 = \{1 - \sum_i (x^i)^2\} dx^2 \wedge dx^3$$

$$-x^3 h\, dx^2 \wedge dh = -x^1 x^3\, dx^1 \wedge dx^2 + (x^3)^2\, dx^2 \wedge dx^3$$

$$x^2 h\, dx^2 \wedge dx^1 = -x^2 \sqrt{1 - \sum_i (x^i)^2}\, dx^1 \wedge dx^2$$

$$-x^2 h\, dh \wedge dx^3 = -x^1 x^2\, dx^3 \wedge dx^1 + (x^2)^2\, dx^2 \wedge dx^3$$

$$x^1 x^2\, dh \wedge dx^2 = -\frac{(x^1)^2 x^2}{\sqrt{1 - \sum_i (x^i)^2}}\, dx^1 \wedge dx^2 + \frac{x^1 x^2 x^3}{\sqrt{1 - \sum_i (x^i)^2}}\, dx^2 \wedge dx^3$$

$$-(x^2)^2\, dh \wedge dx^1 = -\frac{(x^2)^3}{\sqrt{1 - \sum_i (x^i)^2}}\, dx^1 \wedge dx^2 + \frac{(x^2)^2 x^3}{\sqrt{1 - \sum_i (x^i)^2}}\, dx^3 \wedge dx^1$$

$$-x^3 h\, dx^1 \wedge dx^3 = x^3 \sqrt{1 - \sum_i (x^i)^2}\, dx^3 \wedge dx^1$$

$$(x^3)^2\, dx^1 \wedge dh = -\frac{x^2 (x^3)^2}{\sqrt{1 - \sum_i (x^i)^2}}\, dx^1 \wedge dx^2 + \frac{(x^3)^3}{\sqrt{1 - \sum_i (x^i)^2}}\, dx^3 \wedge dx^1$$

$$-x^1 x^3\, dx^3 \wedge dh = \frac{(x^1)^2 x^3}{\sqrt{1 - \sum_i (x^i)^2}}\, dx^3 \wedge dx^1 - \frac{x^1 x^2 x^3}{\sqrt{1 - \sum_i (x^i)^2}}\, dx^2 \wedge dx^3$$

$$-(x^1)^2\, dx^3 \wedge dx^2 = (x^1)^2\, dx^2 \wedge dx^3$$

Thus $$d\omega^1 = \frac{2x^2}{\sqrt{1 - \sum_i (x^i)^2}}\, dx^1 \wedge dx^2 - \frac{2x^3}{\sqrt{1 - \sum_i (x^i)^2}}\, dx^3 \wedge dx^2 - 2\, dx^2 \wedge dx^3.$$

Substituting the above results, one gets

$$\omega_2 \wedge \omega_3 = dx^2 \wedge dx^3 - \frac{x^2}{\sqrt{1 - \sum_i (x^i)^2}}\, dx^1 \wedge dx^2 + \frac{x^3}{\sqrt{1 - \sum_i (x^i)^2}}\, dx^3 \wedge dx^1$$

$$\therefore \quad -2\,\omega_2 \wedge \omega_3 = \frac{2x^2}{\sqrt{1 - \sum_i (x^i)^2}}\, dx^1 \wedge dx^2 - \frac{2x^3}{\sqrt{1 - \sum_i (x^i)^2}}\, dx^3 \wedge dx^1 - 2\, dx^2 \wedge dx^3$$

$$= d\omega_1.$$

Thus $d\omega_1 = -2\,\omega_2 \wedge \omega_3.$

Exercises

Exercise 3.20 *Find the exterior derivative of the following:*

(i) $f = 2xyz^2$ (ii) $f = x^2 + y^2 - 3z^4$ (iii) $f = x^2y + y^2z - z^2x$.

Exercise 3.21 *Find the exterior derivative of:*

(i) $\omega_1 = x^2 y dy - xy^2 dx$.
(ii) $\omega_2 = x^3 dx + yz dy - (x^2 + y^2 + z^2)dz$.
(iii) $\omega_3 = \cos(xy^2)dx \wedge dz$.
(iv) $\omega_4 = xdy \wedge dz + ydz \wedge dx + zdx \wedge dy$.
(v) $\omega_1 \wedge \omega_2$.
(vi) $\omega_1 \wedge \omega_3$.

Exercise 3.22 *Verify (iv) of (3.21) for Exercise 3.21.*

Exercise 3.23 *If $x = r\cos\theta$, $y = r\sin\theta$, then compute $dx \wedge dy$, $dr, d\theta$.*

Exercise 3.24 *From Exercise 3.11(ii), compute $dx \wedge dy \wedge dz$.*

Exercise 3.25 *Consider the following forms in $\mathbb{R}^3$ and verify the property (iii), (iv) of (3.21) where*

(i) $\omega = x^2 dx - z^2 dy$ and $\mu = ydx - xdz$.
(ii) $\omega = xydx + 3dy - yzdz$ and $\mu = xdx - yz^2 dy + 2xdz$.

Answers

3.20. (i) $2yz^2\,dx + 2xz^2\,dy + 4xyz\,dz$.
  (ii) $2x\,dx + 2y\,dy - 12z^3\,dz$.
  (iii) $(2xy - z^2)dx + (x^2 + 2yz)dy + (y^2 - 2zx)dx$.
3.21. (i) $4xydx \wedge dy$ (ii) $-3ydy \wedge dz + 2xdz \wedge dx$ (iii) $2xy\sin(xy^2)dx \wedge dy \wedge dz$.
  (iv) $3dx \wedge dy \wedge dz$ (v) $\{-4xy(x^2 + y^2 + z^2) - 2x^3 y + 3xy^3\}dx \wedge dy \wedge dz$
  (vi) 0.
3.23. $rdr \wedge d\theta$, $\quad dr = \dfrac{xdx + ydy}{\sqrt{x^2 + y^2}}$, $\quad d\theta = \dfrac{xdy - ydx}{\sqrt{x^2 + y^2}}$, $\quad r = \sqrt{x^2 + y^2}$, $\quad \theta = \tan\dfrac{y}{x}$.
3.24. $r^2\sin\theta\,dr \wedge d\theta \wedge d\phi$.

Remark 3.9 It is to be noted that if ϕ is a 0-form, then the 1-form $d\phi$ corresponds to *grad* ϕ. The exterior derivative of a function f corresponds to the gradient vector field.

Verification: In 3-dimension, for a function f,

$$df = \frac{\partial f}{\partial x}dx + \frac{\partial f}{\partial y}dy + \frac{\partial f}{\partial z}dz \leftrightarrow \nabla f = \left(\frac{\partial f}{\partial x}, \frac{\partial f}{\partial y}, \frac{\partial f}{\partial z}\right).$$

Example: Let $\phi = \dfrac{x^2 + y^3}{z}$, then

$$d\phi = \frac{2x}{z}dx + \frac{3y^2}{z}dy - \frac{x^2 + y^3}{z^2}dz$$

$$grad\ \phi = \frac{2x}{z}\vec{i} + \frac{3y^2}{z}\vec{j} - \frac{x^2 + y^3}{z^2}\vec{k}.$$

Remark 3.10 If the 1-form ω corresponds to the vector field V, then the 2-form $d\omega$ corresponds to curl V.

Verification: In 3-dimension, let $\omega = f_1 dx^1 + f_2 dx^2 + f_3 dx^3$ where f_i, $i = 1, 2, 3$ are functions. Then

$$d\omega = df_1 \wedge dx^1 + df_2 \wedge dx^2 + df_3 \wedge dx^3$$

$$= \sum_{i=1}^{3} \frac{\partial f_1}{\partial x^i} dx^i \wedge dx^1 + \sum_{i=1}^{3} \frac{\partial f_2}{\partial x^i} dx^i \wedge dx^2 + \sum_{i=1}^{3} \frac{\partial f_3}{\partial x^i} dx^i \wedge dx^3$$

$$= \left(\frac{\partial f_2}{\partial x^1} - \frac{\partial f_1}{\partial x^2}\right) dx^1 \wedge dx^2 + \left(\frac{\partial f_3}{\partial x^2} - \frac{\partial f_2}{\partial x^3}\right) dx^2 \wedge dx^3 + \left(\frac{\partial f_3}{\partial x^1} - \frac{\partial f_1}{\partial x^3}\right) dx^1 \wedge dx^3$$

$$\equiv \nabla_X f \text{ (Vector Product)}$$

which is basically the **curl**(area).

EXAMPLE:

Let $\omega = (x^2 + y^3 z)dx + (y^2 - 2xz)dy + (x^4 + y^3 - z^2)dz$ in $\mathbb{R}^3$. Then

$$d\omega = (-2z - 3y^2 z)\ dx \wedge dy + (3y^2 + 2x)\ dy \wedge dz + (y^3 - 4x^3)\ dz \wedge dx.$$

Again if $V = (x^2 + y^3 z)\vec{i} + (y^2 - 2xz)\vec{j} + (x^4 + y^3 - z^2)\vec{k}$ is a vector field, then

$$\text{curl } V = (3y^2 + 2x)\vec{i} + (y^3 - 4x^3)\vec{j} + (-2z - 3y^2 z)\vec{k}.$$

Remark 3.11 If the 2-form ω corresponds to the vector field V, then the 3–form $d\omega$ corresponds to the div V.

Verification: In 3-dimension, let ω be a 2-form, where

$\omega = f_{12}\ dx^1 \wedge dx^2 + f_{13}\ dx^1 \wedge dx^3 + f_{23}\ dx^2 \wedge dx^3$, f_{12}, f_{13}, f_{23} being smooth functions. Then

$$d\omega = \sum_{i=1}^{3} \frac{\partial f_{12}}{\partial x^i} dx^i \wedge dx^1 \wedge dx^2 + \sum_{i=1}^{3} \frac{\partial f_{13}}{\partial x^i} dx^i \wedge dx^1 \wedge dx^3 + \sum_{i=1}^{3} \frac{\partial f_{23}}{\partial x^i} dx^i \wedge dx^1 \wedge dx^3$$

$$= \frac{\partial f_{12}}{\partial x^3} dx^3 \wedge dx^1 \wedge dx^2 + \frac{\partial f_{13}}{\partial x^2} dx^2 \wedge dx^1 \wedge dx^3 + \frac{\partial f_{23}}{\partial x^1} dx^1 \wedge dx^2 \wedge dx^3$$

$$= \left(\frac{\partial f_{12}}{\partial x^3} - \frac{\partial f_{13}}{\partial x^2} + \frac{\partial f_{23}}{\partial x^1}\right) dx^1 \wedge dx^2 \wedge dx^3$$

which is the divergence operator.

EXAMPLE:

Let $\omega = (x^2 + y^3 + z^4)\, dy \wedge dz + x^2 y^3 z^4\, dz \wedge dx + (x + 2y + 3z + 1)\, dx \wedge dy$.

Therefore

$$dw = 2x dx \wedge dy \wedge dz + 3x^2 y^2 z^4\, dy \wedge dz \wedge dx + 3 dz \wedge dx \wedge dy$$
$$= (2x + 3x^2 y^2 z^4 + 3)\, dx \wedge dy \wedge dz.$$

If $V = (x^2 + y^3 + z^4)\vec{i} + (x^2 y^3 z^4)\vec{j} + (x + 2y + 3z + 1)\vec{k},$ then

$$\text{div } V = 2x + 3x^2 y^2 z^4 + 3.$$

A form ω is **closed** if

$$d\omega = 0. \tag{3.23}$$

However, if ω is a r-form and

$$d\mu = \omega \tag{3.24}$$

for some $(r - 1)$-form μ, then ω is said to be an **exact form**.

Problem 3.25 *Test whether ω is closed or not, where*

(i) $\omega = xy dx + \left(\dfrac{1}{2}x^2 - y\right) dy.$

(ii) $\omega = e^x \cos y dx + e^x \sin y dy.$

Solution:

(i) Here $d\omega = x dy \wedge dx + \dfrac{1}{2} 2x dx \wedge dy = -x dx \wedge dy + x dx \wedge dy = 0.$

Thus ω is closed.

(ii) Note that $d\omega = -e^x \sin y\, dy \wedge dx + e^x \sin y\, dx \wedge dy = 2e^x \sin y\, dx \wedge dy.$

Thus ω is not closed.

Problem 3.26 *The necessary and sufficient condition that a 1-form ω is a gradient of a function f is that its curl vanishes.*

Solution: If f is a function, then $grad\, f = df$ [refer to Remark 3.9]. Let us assume that 1-form ω be $grad\, f$ i.e. $\omega = grad\, f = df$. By virtue of Remark 3.10, we find

$$curl\, V = d\omega = d(df) = 0 \text{ [refer to (3.21)]},$$

where the 1-form ω corresponds to the vector field V.

For the converse part, suppose for a 1-form ω, $curl\, V = 0$. Taking help of Remark 3.10, we obtain $d\omega = 0$. Then $d^2 f = 0$ holds, for some function f. Thus $\omega = grad\, f$. This completes the proof.

Note that all exact forms are closed but the converse is not always true. The following lemma ensures the converse.

Lemma 3.1 (Poincaré Lemma) *Let ω be a k-form defined on a set*

$$B_r^n = \{(x^1, \ldots, x^n) \in \mathbb{R}^n | (x^1)^2 + (x^2)^2 + \ldots + (x^n)^2 \le r\},$$

such that $d\omega = 0$. Thus $\exists$ a $(k-1)$-form μ defined on B_r^n such that $\omega = d\mu$.

Problem 3.27 Let $\omega = yz\,dx + xz\,dy + xy\,dz$. Find μ such that $d\mu = \omega$.

Solution: Set $\omega = d\mu = \mu_x dx + \mu_y dy + \mu_z dz$, then

$$\int \mu_x dx = \int yz dx = yzx + C_1(y, z), \quad \text{where } C_1 \text{ is a function of } y, z.$$

Then

$$xz = \mu_y = \frac{\partial}{\partial y}(yzx + C_1(y, z)) = zx + C_1'(y, z)$$

$$\Rightarrow C_1'(y, z) = 0$$
$$\Rightarrow C_1(y, z) = \text{constant} = C_2(z)\text{say.}$$

Then $\mu = yzx + C_2(z)$. Now

$$xy = \mu_z = \frac{\partial}{\partial z}(yzx + C_2(z)) = yx + C_2'(z)$$

$$\Rightarrow C_2'(z) = 0$$
$$\Rightarrow C_2(z) = \text{constant} = C\text{say.}$$

Finally, $\mu = yzx + C$.

Alternative

Set $f(x) = \int yz\,dx = yzx + C = \mu(\text{say})$, where C being integration constant, so that $d\mu = d(yzx + C) = zx\,dy + yx\,dz + yz\,dx = \omega$.

Alternative

Set $g(y) = \int xz\,dy = xzy + C = \theta(\text{say})$, where C being integration constant, so that $d\theta = \omega$.

Alternative

Set $h(z) = \int xy\,dz = xyz + C = \phi(\text{say})$, where C being integration constant, so that $d\phi = \omega$.

Problem 3.28 Let $\omega = (12x^2y^3 + 2y)dx \wedge dy$. Find μ such that $d\mu = \omega$.

Solution: Set $f(x, y) = \int (12x^2 y^3 + 2y)dx = 4x^3 y^3 + 2yx + C = \mu(\text{say})$, where C being integration constant, so that

$$
\begin{aligned}
d\mu &= d(4x^3 y^3 + 2yx + C) \wedge dy, \text{ by (3.22)} \\
&= (12x^2 y^3 dx + 12x^3 y^2 dy + 2x dy + 2y dx) \wedge dy \\
&= (12x^2 y^3 + 2y) dx \wedge dy, \text{ as } dy \wedge dy = 0 \\
&= \omega(\text{say}).
\end{aligned}
$$

Problem 3.29 *Compute the exterior derivative of the 2-form*

$$
\omega = \frac{1}{(x^2 + y^2 + z^2)^{3/2}} (x\, dy \wedge dz + y\, dz \wedge dx + z\, dx \wedge dy),
$$

defined on $\mathbb{R}^3 \setminus \{(0, 0, 0)\}$, *where* $(x, y, z) \in \mathbb{R}^3$.

Find the local expression of this form in terms of the spherical coordinates (r, θ, ϕ), *where* $x = r \sin \theta \cos \phi$, $y = r \sin \theta \sin \phi$, $z = r \cos \theta$.

Solution: Note that

$$
\begin{aligned}
d\omega &= d\left\{ \frac{1}{(x^2 + y^2 + z^2)^{3/2}} (x\, dy \wedge dz + y\, dz \wedge dx + z\, dx \wedge dy) \right\} \\
&= d\left(\frac{x}{(x^2 + y^2 + z^2)^{3/2}} \right) \wedge dy \wedge dz + d\left(\frac{y}{(x^2 + y^2 + z^2)^{3/2}} \right) \wedge dz \wedge dx \\
&\quad + d\left(\frac{z}{(x^2 + y^2 + z^2)^{3/2}} \right) \wedge dx \wedge dy, \text{ by (3.22)}.
\end{aligned}
$$

Now

$$
\begin{aligned}
d\left(\frac{x}{(x^2 + y^2 + z^2)^{3/2}} \right) &= d(x(x^2 + y^2 + z^2)^{-3/2}) \\
&= dx(x^2 + y^2 + z^2)^{-3/2} - \frac{3}{2}x(x^2 + y^2 + z^2)^{-5/2}(2x\, dx + 2y\, dy + 2z\, dz) \\
&= \frac{dx}{(x^2 + y^2 + z^2)^{3/2}} - \frac{3x}{(x^2 + y^2 + z^2)^{5/2}}.
\end{aligned}
$$

So, the first term is:

$$
\frac{dx \wedge dy \wedge dz}{(x^2 + y^2 + z^2)^{3/2}} - \frac{3x^2\, dx \wedge dy \wedge dz}{(x^2 + y^2 + z^2)^{5/2}} = \frac{y^2 + z^2 - 2x^2}{(x^2 + y^2 + z^2)^{5/2}} dx \wedge dy \wedge dz.
$$

Similarly, the second term is $\dfrac{z^2 + x^2 - 2y^2}{(x^2 + y^2 + z^2)^{5/2}} dx \wedge dy \wedge dz$ and finally, the third

term is given by $\dfrac{x^2 + y^2 - 2z^2}{(x^2 + y^2 + z^2)^{5/2}} dx \wedge dy \wedge dz$. Hence $d\omega = 0$. Thus ω is closed.

Again

$$x = r \sin \theta \cos \phi, \quad y = r \sin \theta \sin \phi, \quad z = r \cos \theta.$$

Therefore

$$dx = \sin \theta \cos \phi \, dr + r \cos \phi \cos \theta \, d\theta - r \sin \theta \sin \phi \, d\phi,$$
$$dy = \sin \theta \sin \phi \, dr + r \sin \phi \cos \theta \, d\theta + r \sin \theta \cos \phi \, d\phi,$$
$$dz = \cos \theta \, dr - r \sin \theta \, d\theta.$$

After a few steps, one gets

$$xdy \wedge dz = -r^2 \sin \theta \sin \phi \cos \phi \, dr \wedge d\theta + r^2 \sin^2 \theta \cos \theta \cos^2 \phi \, d\phi \wedge dr$$
$$- r^2 \sin^3 \theta \cos^2 \phi \, d\phi \wedge d\theta$$
$$ydz \wedge dx = r^2 \sin \theta \sin \phi \cos \phi \, dr \wedge d\theta - r^2 \sin^2 \theta \sin^2 \phi \cos \theta \, dr \wedge d\phi$$
$$+ r^2 \sin^3 \theta \sin^2 \phi \, d\theta \wedge d\phi$$
$$zdx \wedge dy = r^2 \sin^2 \theta \cos \theta \, dr \wedge d\phi + r^3 \sin \theta \cos^2 \theta \, d\theta \wedge d\phi,$$

where $r^3 = (x^2 + y^2 + z^2)^{3/2}$. Thus the local expression of the given form is $\omega = \dfrac{\sin \theta}{r} d\theta \wedge d\phi$.

Problem 3.30 *Consider the vector fields on* $\mathbb{R}^2$

$$X = x_1 \frac{\partial}{\partial x_1} + 2x_2 \frac{\partial}{\partial x_2}, \quad Y = x_1 x_2 \frac{\partial}{\partial x_2},$$

and let ω *be a* 1-*form on* $\mathbb{R}^2$ *defined by*

$$\omega = (x_1 + x_2^2) \, dx_1 + (x_1^2 + x_2) \, dx_2.$$

Show that ω *satisfies the relation*

$$d\omega(X, Y) = X(\omega(Y)) - Y(\omega(X)) - \omega([X, Y]).$$

Solution: Here

$$\omega(X) = x_1(x_1 + x_2^2) + 2x_2(x_1^2 + x_2) = x_1^2 + x_1 x_2^2 + 2x_1^2 x_2 + 2x_2^2.$$

$$\omega(Y) = x_1 x_2(x_1^2 + x_2) = x_1^3 x_2 + x_1 x_2^2.$$

$$X(\omega(Y)) = 5x_1^3 x_2 + 5x_1 x_2^2.$$

$$Y(\omega(X)) = 2x_1^2 x_2^2 + 2x_1^3 x_2 + 4x_1 x_2^2.$$

$$[X, Y] = \left[x_1 \frac{\partial}{\partial x_1} + 2x_2 \frac{\partial}{\partial x_2}, x_1 x_2 \frac{\partial}{\partial x_2} \right],$$

$$= x_1 x_2 \frac{\partial}{\partial x_2}, \quad \text{after few steps.}$$

Again $\omega([X, Y]) = x_1^3 x_2 + x_1 x_2^2.$

$$\therefore \quad X(\omega(Y)) - Y(\omega(X)) - \omega([X, Y]) = 2x_1^3 x_2 - 2x_1^2 x_2^2.$$

Now $d\omega = 2x_2\, dx_2 \wedge dx_1 + 2x_1\, dx_1 \wedge dx_2,$

$$= (2x_1 - 2x_2)\, dx_1 \wedge dx_2.$$

$$\therefore \quad d\omega(X, Y) = (2x_1 - 2x_2)\, dx_1 \wedge dx_2 \left(x_1 \frac{\partial}{\partial x_1} + 2x_2 \frac{\partial}{\partial x_2}, x_1 x_2 \frac{\partial}{\partial x_2} \right)$$

$$= (2x_1 - 2x_2) \left\{ dx_1 \left(x_1 \frac{\partial}{\partial x_1} + 2x_2 \frac{\partial}{\partial x_2} \right) dx_2 \left(x_1 x_2 \frac{\partial}{\partial x_2} \right) \right.$$

$$\left. - dx_1 \left(x_1 x_2 \frac{\partial}{\partial x_2} \right) dx_2 \left(x_1 \frac{\partial}{\partial x_1} + 2x_2 \frac{\partial}{\partial x_2} \right) \right\}$$

$$= 2x_1^3 x_2 - 2x_1^2 x_2^2.$$

Thus

$$d\omega(X, Y) = X(\omega(Y)) - Y(\omega(X)) - \omega([X, Y]).$$

Exercises

Exercise 3.26 *Test which of the following differential forms are closed and which are exact:*

(i) $\omega_1 = x_2 x_3 dx_1 + x_1 x_3 dy + x_1 x_2 dx_3$

(ii) $\omega_2 = x_2 dx_2 + x_1^2 x_2^2 dy + x_2 x_3 dx_3$

(iii) $\omega_3 = 2xy^2\, dx \wedge dy + z\, dy \wedge dz$

(iv) $\omega_4 = \dfrac{1}{x^2 + y^2}(-y dx + x dy)$ *on* $\mathbb{R}^2 \setminus \{(0, 0)\}.$

Exercise 3.27 *Show that the following forms are exact:*

(i) $\omega = y^2 dx + 2xy dy$

(ii) $\omega = (3y^2 - 4z^4)dx + 6xy dy - 16xz^3 dz$

In each case, find a function f such that $df = \omega$.

Exercise 3.28 *Show that the following 2-forms are exact:*

(i) $\omega = 24x^3 y^2\, dx \wedge dy$

(ii) $\omega = (6x^2 y - 3xy^2)\, dx \wedge dy.$

In each case find a 1-form μ such that $d\mu = \omega$.

Exercise 3.29 *Let* $\omega = zdx \wedge dy - 2xf(x)dy \wedge dz + yf(x)dz \wedge dx$. *Find* f *where*

(i) $d\omega = dx \wedge dy \wedge dz$ *and*
(ii) $d(d\omega) = 0$.

Answers

3.26. (i) closed and exact. (ii) not closed, not exact (iii) closed and exact (iv) closed.

3.27. (i) xy^2 (ii) $3y^2x - 4z^4x$

3.28. (i) $6x^4y^2\,dy$ (ii) $xy^3dx + 2x^3y\,dy$.

3.29. $f(x) = \dfrac{\text{constant}}{\sqrt{x}}$.

Remark 3.12 Given a closed k-form ω, $\nexists$ a $(k-1)$-form μ such that $\omega = d\mu$ i.e. not every closed form is exact. However such an μ always exist **locally** and this result is known as the **Poincaré Lemma**.

A well known example is given by the 1-form $\omega = \dfrac{xdy - ydx}{x^2 + y^2}$ on $M = \mathbb{R}^2 \setminus \{(0,0)\}$, where $d\omega = 0$. If $\gamma : [0, 2\pi] \to M$ is defined by $\gamma(t) = (\cos t, \sin t)$, then

$$\int_\gamma \omega = 2\pi,$$

which is different from zero and thus $\nexists$ a function defined on all of M whose differential coincides with ω.

Remark 3.13 The difference between closed and exact form is measured by **de Rham Cohomology**. The precise definition or computation, of this, is far beyond the scope of this book.

Theorem 3.4 *If ω is a 1-form, then*

$$d\omega(X, Y) = \frac{1}{2}\{X(\omega(Y)) - Y(\omega(X)) - \omega([X, Y])\}, \ \forall\ X, Y \in \chi(M).$$

Proof Without any loss of generality, one may take an 1-form as $\omega = f\,dg$, $\forall\ f, g \in F(M)$. Therefore by virtue of (3.22), we obtain $d\omega = df \wedge dg$. Hence

$$
\begin{aligned}
d\omega(X, Y) &= (df \wedge dg)(X, Y) \\
&= \frac{1}{2}\{df(X)dg(Y) - (df)(Y)dg(X)\} \ \text{by (3.17)} \\
&= \frac{1}{2}\{(Xf)(Yg) - (Yf)(Xg)\} \ \text{by (3.5)}.
\end{aligned}
$$

Using Leibnitz Product Rule, we see that $X(f(Yg)) = (Xf)(Yg) + f(X(Yg))$. Thus

$$dw(X, Y) = \frac{1}{2}\{X(f(Yg)) - f(X(Yg)) - Y(f(Xg)) + f(Y(Xg))\}.$$

Furthermore,

$$\omega(X) = (f dg)(X) = f\, dg(X) = f(Xg), \quad \text{by (3.5)}.$$

Thus the above expression reduces to,

$$dw(X, Y) = \frac{1}{2}\{X(\omega(Y)) - f(X(Yg) - Y(Xg)) - Y(\omega(X))\}$$

$$= \frac{1}{2}\{X(\omega(Y)) - Y(\omega(X)) - f([X, Y]g)\}, \quad \text{by (2.27)}$$

$$= \frac{1}{2}\{X(\omega(Y)) - Y(\omega(X)) - \omega([X, Y])\}, \quad \text{from above}.$$

This completes the proof.

Exercise

Exercise 3.30 *Consider the vector fields X, Y and 1-form ω on $\mathbb{R}^2$ as follows:*

(i) $X = x\dfrac{\partial}{\partial x} + y^2\dfrac{\partial}{\partial y}, \quad Y = y\dfrac{\partial}{\partial y}, \quad \omega = (x + y^2)dx + (y + x^2)dy;$

(ii) $X = 2\dfrac{\partial}{\partial x} - \dfrac{\partial}{\partial y}, \quad Y = e^x\dfrac{\partial}{\partial y} \quad \omega = e^x \cos y\, dx + e^x \sin y\, dy;$

Show that the 1-form ω satisfies Theorem 3.4 in each case.

Theorem 3.5 *If ω is a 1-form, then*

$$dw(fX, Y) = f\, dw(X, Y), \quad f \in F(M), \ \forall\, X, Y \in \chi(M).$$

Proof Using Theorem 3.4, we write

$$dw(fX, Y) = \frac{1}{2}\{(fX)\omega(Y) - Y(\omega(fX)) - \omega([fX, Y])\}.$$

By (2.28), we know that $(fX)\omega(Y) = f(X(\omega(Y)))$.

$$Y(\omega(fX)) = Y(f\omega(X))$$
$$= (Yf)\omega(X) + f(Y(\omega(X))), \quad \text{by Leibnitz Product rule}$$
$$[fX, Y] = f[X, Y] - (Yf)X.$$

Thus taking into consideration (3.1), we have

$$\omega([fX, Y]) = \omega(f[X, Y]) - \omega((Yf)X)$$
$$= f\omega([X, Y]) - (Yf)\omega(X), \ Yf \in F(M).$$

Consequently

$$d\omega(fX, Y) = \frac{1}{2}\{f(X(\omega(Y))) - (Yf)\omega(X) - f(Y(\omega(X))) - f(\omega([X, Y])) + (Yf)\omega(X)\}$$
$$= \frac{1}{2}f\{2d\omega(X, Y)\}, \ \text{refer to Theorem 3.4.}$$
$$\therefore \ d\omega(fX, Y) = f \ d\omega(X, Y).$$

This completes the proof.

EXISTENCE AND UNIQUENESS of Exterior Differentiation:

Without any loss of generality, we may take an r-form as

$$\omega = f_{i_1 i_2 \cdots i_r} \ dx^{i_1} \wedge dx^{i_2} \wedge \cdots \wedge dx^{i_r}, \ \text{where} \ f_{i_1 i_2 \cdots i_r} \in F(M).$$

Let us define an $\mathbb{R}$-linear map $d : \mathfrak{D} \to \mathfrak{D}$ as

$$d\omega = df_{i_1 i_2 \cdots i_r} \wedge dx^{i_1} \wedge dx^{i_2} \wedge \cdots \wedge dx^{i_r}.$$

Then

(i) $d(\mathfrak{D}_r) \subset \mathfrak{D}_{r+1}$
(ii) if ω is a 0-form, then $d\omega$ is the total differential of ω
(iii) let $\mu \in \mathfrak{D}_s$. We consider $\mu = g_{j_1 j_2 \cdots j_s} \ dx^{j_1} \wedge dx^{j_2} \wedge \cdots \wedge dx^{j_s}$, where $g_{j_1 j_2 \cdots j_s} \in F(M)$. Then we get

$$\omega \wedge \mu = f_{i_1 i_2 \cdots i_r} g_{j_1 j_2 \cdots j_s} \ dx^{i_1} \wedge dx^{i_2} \wedge \cdots \wedge dx^{i_r} \wedge dx^{j_1} \wedge dx^{j_2} \wedge \cdots \wedge dx^{j_s}.$$

Using (3.22), we find

$$d(\omega \wedge \mu) = d(f_{i_1 i_2 \cdots i_r} g_{j_1 j_2 \cdots j_s}) \wedge dx^{i_1} \wedge dx^{i_2} \wedge \cdots \wedge dx^{i_r} \wedge dx^{j_1} \wedge dx^{j_2} \wedge \cdots \wedge dx^{j_s}$$
$$= g_{j_1 j_2 \cdots j_s} \ df_{i_1 i_2 \cdots i_r} \wedge dx^{i_1} \wedge dx^{i_2} \wedge \cdots \wedge dx^{i_r} \wedge dx^{j_1} \wedge dx^{j_2} \wedge \cdots \wedge dx^{j_s}$$
$$+ f_{i_1 i_2 \cdots i_r} \ dg_{j_1 j_2 \cdots j_s} \wedge dx^{i_1} \wedge dx^{i_2} \wedge \cdots \wedge dx^{i_r} \wedge dx^{j_1} \wedge dx^{j_2} \wedge \cdots \wedge dx^{j_s}$$
$$= df_{i_1 i_2 \cdots i_r} \wedge dx^{i_1} \wedge dx^{i_2} \wedge \cdots \wedge dx^{i_r} \wedge g_{j_1 j_2 \cdots j_s} \ dx^{j_1} \wedge dx^{j_2} \wedge \cdots \wedge dx^{j_s} +$$
$$+ (-1)^r f_{i_1 i_2 \cdots i_r} \ dx^{i_1} \wedge dx^{i_2} \wedge \cdots \wedge dx^{i_r} \wedge dg_{j_1 j_2 \cdots j_s} \wedge dx^{j_1} \wedge dx^{j_2} \wedge \cdots \wedge dx^{j_s}$$
$$= d\omega \wedge \mu + (-1)^r \omega \wedge d\mu.$$

(iv) Finally, from (3.22) we obtain

$$dw = \sum_{i_s} \frac{\partial f}{\partial x^{i_s}} dx^{i_s} \wedge dx^{i_1} \wedge \cdots \wedge dx^{i_r}$$

$$\text{or, } d(dw) = \sum_{i_k, i_s} \frac{\partial^2 f}{\partial x^{i_k} \partial x^{i_s}} dx^{i_k} \wedge dx^{i_s} \wedge dx^{i_1} \wedge \cdots \wedge dx^{i_r}$$

$$= 0.$$

Thus d satisfies (3.21) and hence the existence is established. It is easy to establish the uniqueness of d. Consequently, there exists a unique exterior differentiation on $\mathfrak{D}$.

Exercise 3.31 *If ω is a 1-form on M, prove that*

$$d\omega(X, fY) = f \, d\omega(X, Y), \quad f \in F(M), \ \forall \, X, Y \in \chi(M).$$

3.4 Pull-Back Differential Form

Let M be an n-dimensional and N be an m-dimensional manifold and $f : M \to N$ be a differentiable mapping (Fig. 3.1).

Let $T_p(M)$ be the tangent space at $p \in M$ and $T_p^*(M)$ be its dual. Let $T_{f(p)}(N)$ and $T_{f(p)}^*(N)$ be respectively the tangent space and dual space at $f(p) \in N$.

Let ω be an 1-form on $\mathfrak{D}_1(N) \equiv T_{f(p)}^*(N)$. We define an 1-form on $\mathfrak{D}_1(M) \equiv T_p^*(M)$ called the **pull-back** 1-**form** at $p \in M$, denoted by $f^*\omega$ as follows:

$$\{f^*(\omega_{f(p)})\}(X_p) = \omega_{f(p)}\{f_*(X_p)\}, \ \forall \, p \in M \qquad (3.25)$$

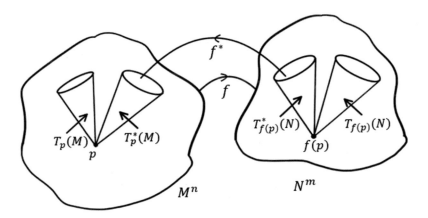

Fig. 3.1 Pull-Back Differential Form

where f^*, f_* are already defined in §2.12.

Let us write

$$f^*(\omega_{f(p)}) = (f^*\omega)_p. \tag{3.26}$$

Then (3.25) reduces to

$$(f^*\omega)_p X_p = \omega_{f(p)}(f_*X)_{f(p)} \quad \text{by (2.35).} \tag{3.27}$$

We write it as

$$(f^*\omega)(X) = \omega(f_*X). \tag{3.28}$$

Problem 3.31 Let $f : M^n \to N^m$ be a C^∞ map between two manifolds. Let ω be a C^∞ r-form on N. Show that $f^*\omega$ is also a C^∞ r-form on M.

Solution: Let $(U, x^1, \ldots, x^n)$ and $(V, y^1, \ldots, y^m)$ be two charts on M and N respectively, where $p \in U \subset M$ and $f(p) \in V \subset N$. As ω is a C^∞ r-form on N, by Theorem 3.2

$$\omega = \sum_{\substack{j_1, j_2, \ldots, j_r \\ j_1 < j_2 < \ldots < j_r}} g_{j_1 j_2 \cdots j_m} \, dy^{j_1} \wedge dy^{j_2} \wedge \ldots \wedge dy^{j_r},$$

where $g_{j_1 j_2 \cdots j_m}$ are C^∞ functions on N such that $g_{j_1 j_2 \cdots j_m} \in F(N)$. Then from (3.33), we have

$$f^*\omega = \sum_{\substack{j_1, j_2, \ldots, j_r \\ j_1 < j_2 < \ldots < j_r}} (g_{j_1 j_2 \cdots j_r} \circ f) \, df^{j_1} \wedge df^{j_2} \wedge \ldots \wedge df^{j_r}.$$

By virtue of Theorem 3.3, one gets

$$f^*\omega = \sum_{\substack{i_1, i_2, \ldots, i_r \\ i_1 < i_2 < \ldots < i_r}} (g_{j_1 j_2 \cdots j_r} \circ f) \frac{\partial(f^{j_1}, f^{j_2}, \ldots, f^{j_r})}{\partial(x^{i_1}, x^{i_2}, \ldots, x^{i_r})} \, dx^{i_1} \wedge dx^{i_2} \wedge \ldots \wedge dx^{i_r}.$$

Now $(g_{j_1 j_2 \cdots j_r} \circ f)$ and $\dfrac{\partial(f^{j_1}, f^{j_2}, \ldots, f^{j_r})}{\partial(x^{i_1}, x^{i_2}, \ldots, x^{i_r})}$ are all C^∞ functions. Thus, $f^*\omega$ is also a C^∞ r-form on M.

Theorem 3.6 If f is a differentiable mapping from an n-dimensional manifold M to an m-dimensional manifold N where $(x^1, x^2, \ldots, x^n)$ and $(y^1, y^2, \ldots, y^m)$ are respectively the coordinates at $p \in M$ and $f(p) \in N$, then

$$f^*(dy^j)_{f(p)} = \sum_{i=1}^{n} \left(\frac{\partial f^j}{\partial x^i}\right)_p (dx^i)_p, \quad f^j = y^j \circ f, \; j = 1, 2, 3, \ldots, m.$$

Proof We write

$$f^*(dy^j)_{f(p)} = \sum_{i=1}^{n} a_i^j (dx^i)_p,$$

where a_i^j's are to be determined. Therefore

$$f^*(dy^j)_{f(p)} \left(\frac{\partial}{\partial x^k} \right)_p = a_k^j, \text{ by } (3.6)$$

$$\text{or } (dy^j)_{f(p)} \left\{ f_* \left(\frac{\partial}{\partial x^k} \right)_p \right\} = a_k^j, \text{ by } (3.25)$$

$$\text{or } (dy^j)_{f(p)} \sum_{s=1}^{m} \left(\frac{\partial f^s}{\partial x^k} \right)_p \left(\frac{\partial}{\partial y^s} \right) = a_k^j, \text{ by Theorem 2.6 of } Sect. 2.12 \text{ and } f^s = y^s \circ f.$$

$$\therefore \quad \frac{\partial f^j}{\partial x^k} = a_k^j, \text{ by } (3.6).$$

Consequently, we have

$$f^*(dy^j)_{f(p)} = \sum_{i=1}^{n} \frac{\partial f^j}{\partial x^i} (dx^i)_p, \quad j = 1, 2, 3, \ldots, m.$$

Remark 3.14 From Theorem 3.6, we observe

$$f^*(dy^j)_{f(p)} = (df^j)_p, \quad j = 1, 2, 3, \ldots, m. \tag{3.29}$$

Remark 3.15 We can write

$$f^*(dy^j)_{f(p)} = d(y^j \circ f)_p, \quad j = 1, 2, 3, \ldots, m. \tag{3.30}$$

Theorem 3.7 Let $f : M^n \to N^m$ be a C^∞ map. If ω, μ are 1-forms on N, then

(i) $f^*(\omega + \mu) = f^*\omega + f^*\mu$
(ii) $f^*(g\omega) = f^*(g)f^*(\omega)$
(iii) $f^*(\lambda\omega) = \lambda f^*\omega, \quad \forall g \in F(N), \lambda \in \mathbb{R}.$

Proof (i) Since $\omega + \mu$ is also a 1-form, in view of (3.25), one gets

$$\{f^*(\omega + \mu)_{f(p)}\}(X_p) = (\omega + \mu)_{f(p)}\{f_*(X_p)\}$$
$$= \omega_{f(p)}\{f_*(X_p) + \mu_{f(p)}\{f_*(X_p)\}.$$

Furthermore, applying (3.25) on the right hand side of the foregoing equation, the result follows immediately.

(ii) Again for $g \in F(N)$, $\omega \in \mathfrak{D}_1(N)$, $g\omega \in \mathfrak{D}_1(N)$ and hence from (3.25), we have

$$\{f^*(g\omega)_{f(p)}\}(X_p) = (g\omega)_{f(p)}\{f_*(X_p)\}$$
$$= (g \circ f)(p)\omega_{f(p)}\{f_*(X_p)\}, \text{ by } (3.12)$$
$$\text{or } \{f^*(g\omega)\}_p(X_p) = (g \circ f)(p)(f^*\omega)_p(X_p), \text{ by } (3.26), (3.27)$$
$$= (f^*g)(p)(f^*\omega)_p(X_p), \text{ by } (2.32)$$
$$f^*(g\omega) = f^*(g)f^*(\omega), \ \forall \ p \in M.$$

(iii) Left to the reader.

Let $\omega \in \mathfrak{D}_r(N)$. Then a **pull-back r-form** on M, denoted by $f^*\omega$, is defined as follows:

$$\{f^*(\omega)_{f(p)}\}\{(X_1)_p, (X_2)_p, \ldots, (X_r)_p\} = \omega_{f(p)}\{f_*(X_1)_p), f_*(X_2)_p), \ldots, f_*(X_r)_p)\}. \tag{3.31}$$

We also write it as

$$(f^*\omega)(X_1, X_2, \ldots, X_r) = \omega(f_*X_1, f_*X_2, \ldots, f_*X_r). \tag{3.32}$$

For any $\omega \in \mathfrak{D}_r(N)$, by (3.19) we can write

$$\omega = \sum_{i_1,i_2,\ldots,i_r} g_{i_1,i_2,\ldots,i_r} dx^{i_1} \wedge dx^{i_2} \wedge dx^{i_3} \wedge \cdots \wedge dx^{i_r}, \ i_1 < i_2 < \cdots < i_r.$$

Taking into consideration Theorem 3.7(ii), we have

$$f^*\omega = \sum f^*(g_{i_1,i_2,\ldots,i_r})f^*(dx^{i_1}) \wedge f^*(dx^{i_2}) \wedge \cdots \wedge f^*(dx^{i_r}). \tag{3.33}$$

Combining (3.29) and (2.32), we find

$$f^*\omega = \sum_{i_1,i_2,\ldots,i_r} (g_{i_1,i_2,\ldots,i_r} \circ f)\, df^{i_1} \wedge df^{i_2} \wedge df^{i_3} \wedge \cdots \wedge df^{i_r}, \ i_1 < i_2 < \cdots < i_r. \tag{3.34}$$

Theorem 3.8 *If $f : M^n \to N^m$ is a C^∞ map and ω, μ are any forms, then*

(i) $f^*(\omega \wedge \mu) = f^*\omega \wedge f^*\mu$.
(ii) $(f \circ h)^*\omega = h^*(f^*\omega)$, h *being a C^∞ function.*

Proof

(i) Let $\omega \in \mathfrak{D}_r(N)$ and $\mu \in \mathfrak{D}_r(N)$. Then $\omega \wedge \mu \in \mathfrak{D}_{r+s}(N)$ and hence by (3.32), we get

$$\{f^*(\omega \wedge \mu)(X_1, X_2, \ldots, X_r, Y_1, Y_2, \ldots, Y_s)\}$$
$$= (\omega \wedge \mu)(f_*X_1, \ldots, f_*X_r; f_*Y_1, \ldots, f_*Y_s)$$
$$= \frac{1}{(r+s)!}\sum_\sigma (\text{sgn } \sigma)\omega(f_*X_{\sigma(1)}, \ldots, f_*X_{\sigma(r)})\mu(f_*Y_{\sigma(1)}, \ldots, f_*Y_{\sigma(s)}), \text{ by } (3.14).$$

The result follows immediately.

(ii) Using (3.30) and (2.32), we can write (3.33) as

$$f^*\omega = \sum_i (g_{i_1,i_2,\dots,i_r} \circ f) d(x^{i_1} \circ f) \wedge d(x^{i_2} \circ f) \wedge \cdots \wedge d(x^{i_r} \circ f).$$

$$\text{or } h^*(f^*\omega) = \sum_i \{(g_{i_1,i_2,\dots,i_r} \circ f) \circ h\} d((x^{i_1} \circ f) \circ h) \wedge \cdots \wedge d((x^{i_r} \circ f) \circ h).$$

Using Theorem 3.7(ii) and (3.19), we obtain

$$(f \circ h)^* \left(\sum_i g_{i_1,i_2,\dots,i_r} dx^{i_1} \wedge dx^{i_2} \wedge \cdots \wedge dx^{i_r} \right)$$

$$= \sum_i ((f \circ h)^* g_{i_1,i_2,\dots,i_r})(f \circ h)^* dx^{i_1} \wedge (f \circ h)^* dx^{i_2} \wedge \cdots \wedge (f \circ h)^* dx^{i_r}$$

$$= \sum_i (g_{i_1,i_2,\dots,i_r} \circ (f \circ h)) d(x^{i_1} \circ (f \circ h)) \wedge d(x^{i_2} \circ (f \circ h)) \wedge \cdots \wedge d(x^{i_r} \circ (f \circ h)), \text{ by (3.30)}.$$

Using the associativity of C^∞ functions, the result follows.

Theorem 3.9 *For any form ω, $d(f^*\omega) = f^*(d\omega)$.*

Proof The following cases will arise:

Case (I) ω **is a 0-form:**
Let $\omega = h$, h being a C^∞ function.

$$\{f^*(dh)\}(X) = dh(f_* X), \text{ by (3.25)}$$
$$= (f_* X)h, \text{ by (3.5)}$$
$$= X(h \circ f), \text{ by (2.34)}$$
$$= (d(h \circ f))X, \text{ by (3.5)}$$
$$= d(f^*h)X, \text{ by (2.32)}.$$

Thus $d(f^*\omega) = f^*(d\omega)$ holds for every X.

Case (II) ω **is a r-form:**
Taking aid of Principle of Mathematical Induction, we assume that the result is true for $(r-1)$-form. Without any loss of generality, any r-form can be expressed as

$$\omega = g_{i_1,i_2,\dots,i_r} dx^{i_1} \wedge dx^{i_2} \wedge dx^{i_3} \wedge \cdots \wedge dx^{i_r}.$$

Using Theorem 3.8(i), we have

$$f^*\omega = f^*(g_{i_1,i_2,\dots,i_r} dx^{i_1} \wedge dx^{i_2} \wedge dx^{i_3} \wedge \cdots \wedge dx^{i_{r-1}}) \wedge f^*(dx^{i_r})$$
$$\text{or } d(f^*\omega) = d\{f^*(g_{i_1,i_2,\dots,i_r} dx^{i_1} \wedge dx^{i_2} \wedge dx^{i_3} \wedge \cdots \wedge dx^{i_{r-1}}) \wedge f^*(dx^{i_r})\}.$$

In view of (3.21)(iii), we get

$$d(f^*\omega) = d\{f^*(g_{i_1, i_2, \ldots, i_r} dx^{i_1} \wedge dx^{i_2} \wedge dx^{i_3} \wedge \cdots \wedge dx^{i_{r-1}})\} \wedge f^*(dx^{i_r}) +$$
$$+ (-1)^{r-1} f^*(g_{i_1, i_2, \ldots, i_r} dx^{i_1} \wedge dx^{i_2} \wedge dx^{i_3} \wedge \cdots \wedge dx^{i_{r-1}}) d(f^*(dx^{i_r})).$$

Note that dx^{i_r} is a 1-form and as the result is true for $(r-1)$ form,

$$d(f^*(dx^{i_r})) = f^*(d(dx^{i_r})) = 0, \text{ by (3.21)(iv).}$$

Thus

$$d(f^*\omega) = d\{f^*(g_{i_1, i_2, \ldots, i_r} dx^{i_1} \wedge dx^{i_2} \wedge dx^{i_3} \wedge \cdots \wedge dx^{i_{r-1}})\} \wedge f^*(dx^{i_r})\}$$
$$= f^*\{d(g_{i_1, i_2, \ldots, i_r} dx^{i_1} \wedge dx^{i_2} \wedge dx^{i_3} \wedge \cdots \wedge dx^{i_{r-1}})\} \wedge f^*(dx^{i_r}),$$
$$\text{as it is true for } (r-1) - \text{form}$$
$$= f^*\{dg_{i_1, i_2, \ldots, i_r} \wedge dx^{i_1} \wedge dx^{i_2} \wedge dx^{i_3} \wedge \cdots \wedge dx^{i_{r-1}}\} \wedge f^*(dx^{i_r}), \text{ by (3.22)}$$
$$= f^*\{dg_{i_1, i_2, \ldots, i_r} \wedge dx^{i_1} \wedge dx^{i_2} \wedge dx^{i_3} \wedge \cdots \wedge dx^{i_{r-1}} \wedge dx^{i_r}, \text{ refer to Theorem 3.8(i)}$$
$$= f^*(d\omega), \text{ by (3.22).}$$

Thus the result is true for r-form. Hence, we can claim that for any form ω

$$d(f^*\omega) = f^*(d\omega).$$

Problem 3.32 Let $g = x^2 - y^2$, $\omega = xydx + xdy$. Let $f : \mathbb{R} \to \mathbb{R}^2$ be given by $f(t) = (t, t^2) = (x, y)$. Find **A. (i)** f^*g **(ii)** $f^*\omega$. **B.** Show that $f^*(dg) = d(f^*g)$.

Solution:

A. (i) $f^*g = g \circ f = t^2 - t^4$.
  (ii) Here

$$\begin{aligned}
f^*\omega &= f^*(xydx + xdy) \\
&= f^*(xydx) + f^*(xdy), \text{ by linearity} \\
&= f^*(xy)f^*(dx) + f^*(x)f^*(dy), \text{ by Theorem 3.7(ii)} \\
&= (xy \circ f)d(f^*x) + (x \circ f)d(f^*y), \text{ by (2.32)} \\
&= (t \circ t^2)d(x \circ f) + td(y \circ f), \text{ by (3.30)} \\
&= t^3 d(t) + td(t^2) \\
&= t^3 dt + t(2tdt) \\
&= (t^3 + 2t^2)dt.
\end{aligned}$$

ALTERNATIVE METHOD: $f^*\omega = f^*(xydx + xdy) = (t \circ t^2)dt + td(t^2) = (t^3 + 2t^2)dt$.

B. Note that

$$d(f^*g) = d(t^2 - t^4) = (2t - 4t^3)dt,$$
$$dg = d(x^2 - y^2) = 2xdx - 2ydy$$
$$f^*(dg) = f^*(2xdx - 2ydy) = 2tdt - 2t^2(2tdt) = (2t - 4t^3)dt.$$

Thus $f^*(dg) = d(f^*g)$.

Problem 3.33 *Let U be the open set* $(0, \infty) \times (0, 2\pi)$ *in the* (ρ, θ)-*plane* R^2. *Define* $f : U \subset R^2 \to R^2$ *by* $f(\rho, \theta) = (\rho \cos \theta, \rho \sin \theta)$. *If x, y are the standard coordinates on* R^2, *compute the pull-back* $f^*(dx \wedge dy)$.

Solution: We know that

$$f^*(dx \wedge dy) = f^*dx \wedge f^*dy, \text{ where}$$
$$f^*dx = d(\rho \cos \theta) = \cos \theta d\rho - \rho \sin \theta d\theta, \text{ and}$$
$$f^*dy = d(\rho \sin \theta) = \sin \theta d\rho + \rho \cos \theta d\theta.$$
$$\therefore \ f^*(dx \wedge dy) = (\cos \theta d\rho - \rho \sin \theta d\theta) \wedge (\sin \theta d\rho + \rho \cos \theta d\theta)$$
$$= \rho \cos^2 \theta d\rho \wedge d\theta + \rho \sin^2 \theta d\rho \wedge d\theta$$
$$= \rho d\rho \wedge d\theta.$$

Problem 3.34 *Consider a map* $f : U \subset R^4 \to R^2$ *given by* $f(x_1, x_2, x_3, x_4) = (u, v)$ *where*

$$u = x_1^2 + x_2^2 + x_3^2 + x_4^2 - 1, \quad v = x_1^2 + x_2^2 + x_3^2 + x_4^2 - 2x_2 - 2x_3 + 5.$$

Calculate $f^*_{(-1,5)} (du + 2dv) \in T^*_{(-1,5)} (R^2)$, *taking* $(0, 0, 0, 0)$ *at* $f^{-1}(-1, 5)$.

Solution: Consider $x_1 = x_2 = x_3 = x_4 = 0, \ u = -1, v = 5$. Now

$$f^*(du + 2dv) = d(x_1^2 + x_2^2 + x_3^2 + x_4^2 - 1) + 2d(x_1^2 + x_2^2 + x_3^2 + x_4^2 - 2x_2 - 2x_3 + 5)$$
$$= 6x_1 dx_1 + 6x_2 dx_2 + 6x_3 dx_3 + 6x_4 dx_4 - 4dx_2 - 4dx_3.$$
$$\therefore \ f^*_{(-1,5)} (du + 2dv) = (-4dx_2 - 4dx_3)_{(0,0,0,0)}$$
$$= -4(dx_2 + dx_3)_{(0,0,0,0)}.$$

Problem 3.35 *Let M be a circle and N be* R^2 *so that the map* $f : M \to N$ *be defined by* $x^1 = r \cos \theta, x^2 = r \sin \theta$. *If* $\omega = adx^1 + bdx^2$ *and* $\mu = \frac{1}{a}dx^1 + \frac{1}{b}dx^2$, *compute* $f^*(\omega \wedge \mu)$, *where a, b are constants.*

Solution: From Theorem 3.8(i), we have

$$f^*(\omega \wedge \mu) = f^*\omega \wedge f^*\mu.$$

Now

$$
\begin{aligned}
f^*\omega &= f^*(adx^1 + bdx^2) = af^*(dx^1) + bf^*(dx^2) \\
&= ad(f^*x^1) + bd(f^*x^2), \quad \text{by Theorem 3.8} \\
&= ad(x^1 \circ f) + bd(x^2 \circ f) \\
&= ad(r \cos \theta) + b(r \sin \theta) \\
&= a(\cos \theta \, dr - r \sin \theta \, d\theta) + b(\sin \theta \, dr + r \cos \theta \, d\theta) \\
&= (a \cos \theta + b \sin \theta) \, dr + (br \cos \theta - ar \sin \theta) \, d\theta
\end{aligned}
$$

Similarly

$$f^*\mu = \left(\frac{1}{a} \cos \theta + \frac{1}{b} \sin \theta\right) dr + \left(\frac{r}{b} \cos \theta - \frac{r}{a} \sin \theta\right) d\theta.$$

$$
\begin{aligned}
\therefore \quad f^*(\omega \wedge \mu) &= \left\{(a \cos \theta + b \sin \theta)\left(\frac{r}{b} \cos \theta - \frac{r}{a} \sin \theta\right) dr \wedge d\theta\right\} \\
&\quad + \left\{(br \cos \theta - ar \sin \theta)\left(\frac{1}{a} \cos \theta + \frac{1}{b} \sin \theta\right)\right\} d\theta \wedge dr \\
&= r\left(\frac{a}{b} - \frac{b}{a}\right) dr \wedge d\theta, \quad \text{after a few steps.}
\end{aligned}
$$

Problem 3.36 *Let U be the open set $(0, \infty) \times (0, \pi) \times (0, 2\pi)$ in the (r, ϕ, θ)-space $\mathbb{R}^3$. Let $f : U \to \mathbb{R}^3$ be defined by*

$$f(r, \phi, \theta) = (r \sin \phi \cos \theta, r \sin \phi \sin \theta, r \cos \phi).$$

If (x, y, z) are the standard coordinates of $\mathbb{R}^3$, show that

$$f^*(dx \wedge dy \wedge dz) = r^2 \sin \phi \, dr \wedge d\phi \wedge d\theta.$$

Solution: It is known that

$$f^*(dx \wedge dy \wedge dz) = f^*dx \wedge f^*dy \wedge f^*dz.$$

Now

$$f^* dx = d(f^* x) = d(x \circ f) = df^1 = d(r \sin \phi \cos \theta)$$
$$= dr \sin \phi \cos \theta + r \cos \phi \cos \theta \, d\phi - r \sin \phi \sin \theta \, d\theta.$$

Similarly

$$f^* dy = \sin \phi \sin \theta \, dr + r \cos \phi \sin \theta \, d\phi + r \sin \phi \cos \theta \, d\theta,$$
$$f^* dz = \cos \phi \, dr - r \sin \phi \, d\phi.$$

Therefore

$$f^* (dx \wedge dy \wedge dz) = \{ (\sin \phi \cos \phi \sin \theta \cos \theta) \, dr \wedge d\phi + (r \sin^2 \phi \cos^2 \theta) \, dr \wedge d\theta$$
$$+ (r \sin \phi \cos \phi \sin \theta \cos \theta) \, d\phi \wedge dr + (r^2 \sin \phi \cos \phi \cos^2 \theta) \, d\phi \wedge d\theta$$
$$- (r \sin^2 \phi \sin^2 \theta) \, d\theta \wedge dr - (r^2 \sin \phi \cos \phi \sin^2 \theta) \, d\theta \wedge d\phi \}$$
$$\wedge (\cos \phi \, dr - r \sin \phi \, d\phi)$$
$$= (-r^2 \sin^3 \phi \cos^2 \theta) \, dr \wedge d\theta \wedge d\phi + (r^2 \sin \phi \cos^2 \phi \cos^2 \theta) \, d\phi \wedge d\theta \wedge dr$$
$$+ (r^2 \sin^3 \phi \sin^2 \theta) \, d\theta \wedge dr \wedge d\phi - (r^2 \sin \phi \cos^2 \phi \sin^2 \theta) \, d\theta \wedge d\phi \wedge dr$$
$$= r^2 \sin \phi \, dr \wedge d\phi \wedge d\theta.$$

Exercises

Exercise 3.32 Let $h = 2xy$, $\mu = -2y dx + x dy$, $\theta = (x^2 + y^2) dy$. Let $f : \mathbb{R} \to \mathbb{R}^2$ be given by $f(t) = (t, t^2) = (x, y)$.

A. Find (i) $f^* h$ (ii) $f^* \mu$ (iii) $f^* \theta$.
B. Show that $f^*(dh) = d(f^* h)$.

Exercise 3.33 Let $f : \mathbb{R}^2 \to \mathbb{R}^2$ be given by

$$f(u, v) = (u^2 + 1, uv) = (x, y), \quad \omega_1 = (xy - y) dx \wedge dy, \quad \mu_1 = (x + y^2) dx \wedge dy.$$

Find (i) $f^* h$ (ii) $f^* \mu$ (iii) $f^* \theta$ (iv) $f^* \omega_1$ (v) $f^* \mu_1$, where h, μ, θ are defined in Exercise 3.32.

Exercise 3.34 Let $\mu = -2y dx + x dy$, $\theta = (x^2 + y^2) dy$. Let $f : \mathbb{R}^2 \to \mathbb{R}^2$ be given by $f(u, v) = (u^2 + 1, uv) = (x, y)$. Show that

(i) $f^*(\mu \wedge \theta) = f^* \mu \wedge f^* \theta$.
(ii) $f^*(\omega \wedge \theta) = f^* \omega \wedge f^* \theta$, where $\omega = xy dx + x dy$.

Exercise 3.35 If $f : M \to \mathbb{R}^3$ be such that $f(u, v) = (u \cos v, u \sin v, av)$, then for a given 1-form $\omega = x^1 dx^1 - dx^2 + x^2 dx^3$ on $\mathbb{R}^3$, where $x^1 = u \cos v, x^2 = u \sin v, x^3 = av$, compute $f^* \omega$.

Exercise 3.36 *If $f : M \to \mathbb{R}^3$ be such that $f(u, v) = (a \cos u \sin v, a \sin u \sin v, a \cos v)$, then for a given 1-form $\omega = dx^1 + a dx^2 + a dx^3$ on $\mathbb{R}^3$, determine $f^*\omega$.*

Exercise 3.37 *Let $\omega = -\dfrac{y}{x^2 + y^2} dx + \dfrac{x}{x^2 + y^2}$ be the 1-form in $\mathbb{R}^2 \setminus \{(0, 0)\}$. Let $U = \{r > 0 : 0 < \theta < 2\pi\}$ be the set in the plane (r, θ) and $f : U \to \mathbb{R}^2$ be the map $f(r, \theta) = \begin{cases} x = r \cos \theta \\ y = r \sin \theta \end{cases}$. Compute $f^*\omega$.*

Exercise 3.38 *Let S^1 be a unit circle and $f : \mathbb{R} \to S^1 \subset \mathbb{R}^2$ be given by $f(\theta) = (x_1, x_2) = (\cos \theta, \sin \theta)$. If ω is the 1-form $-x_2 dx_1 + x_1 dx_2$ on S^1, compute $f^*\omega$.*

Exercise 3.39 *Consider on $\mathbb{R}^2$, $\theta = (2xy + x^2 + 1)dx + (x^2 - y)dy$ and f be the map $f : \mathbb{R}^3 \to \mathbb{R}^2$ given by $(u, v, w) \mapsto (x, y) = (u - v, v^2 + w)$. Compute $f^*\theta$.*

Exercise 3.40 *Let $f : \mathbb{R}^2 \to \mathbb{R}^2$ be given by $f(x_1, x_2) = (u, v)$, where $u = x_1^2 + x_2^2$, $v = x_1 x_2$. Calculate*

(i) $f^*(udu + vdv)$ *and*
(ii) $f^*_{(2,1)}(udu + vdv) \in T^*_{(2,1)}(\mathbb{R}^2)$, *taking $(1, 1)$ to $f^{-1}(2, 1)$.*

Answers

3.32. (i) $2t^3$ (ii) 0   (iii) $(2t^3 + 2t^5)dt$

3.33. (i) $2u^3v + 2uv$ (ii) $(-3u^2v + v)du + (u^3 + u)dv$
(iii) $(u^4v + 2u^2v + v + u^2v^3)du + (u^5 + 2u^3 + u + u^3v^2)dv$
(iv) $2u^5v\, du \wedge dv$ (v) $(2u^4 + 2u^2 + 2u^4v^2)du \wedge dv$

3.35. $(u \cos^2 v - \sin v)du + (au \sin v - u \cos v - u^2 \sin v \cos v)dv$

3.36. $(-a \sin u \sin v + a^2 \sin v \cos u)du + (a \cos u \cos v + a^2 \sin u \cos v - a \sin v)dv$

3.37. $d\theta$ **3.38.** $d\theta$

3.39. $\{2(u - v)(v^2 + w) + (u - v)^2 + 1\}du + (u^2 - 2uv - w)d\omega$
$+(-4uv - 2u\omega + 2v^3 - u^2 + 2uv - v^2 - 1 + 2u^2v)dv.$

3.40. (i) $(2x_1^3 + 3x_1 x_2^2)dx_1 + (2x_2^3 + 3x_1^2 x_2)dx_2$.   (ii) $5(dx_1 + dx_2)\big|_{(1,1)}$.

Chapter 4
Lie Group

4.1 Lie Group, Left and Right Translation

A Lie group is a group (in algebraic sense), which is also a differentiable manifold, with the property that the group operations are compatible with the smooth structure. Thus, a Lie group G consists of two structures on the same set G, namely, it is a differentiable manifold and has also a group structure. We now state the formal definition as follows:

Let G be a non-empty set. If

$$\begin{cases} (i) & G \text{ is a group (whose operation is denoted by multiplication),} \\ (ii) & G \text{ is an } n-\text{dimensional smooth manifold and} \\ (iii) & \text{the inverse map } \tau : G \to G \text{ such that } \tau(x) = x^{-1} \text{ and} \\ & \text{the multiplication map } \phi : G \times G \to G \text{ such that } \phi(x, y) = xy, \ \forall\, x, y \\ & \text{are smooth maps, then } G \text{ is called an } n\text{-\textbf{dimensional Lie Group}.} \end{cases}$$

$$(4.1)$$

Remark 4.1 The group is called "Lie Group", after the Norwegian mathematician Sophus Lie (1842–1899).

Remark 4.2 (a) The product of two second countable and Hausdorff spaces is respectively the second countable and Hausdorff space. [for details refer to any standard textbook of topology]

(b) If M and N are C^∞ manifolds, then $M \times N$ with its product topology is Hausdorff and second countable. To show that $M \times N$ is a smooth manifold, it remains to find a C^∞ atlas. If we have a chart (U, ϕ_U) and (V, ϕ_V) respectively on M and N, then $U \times V \subseteq M \times N$ an open subset of the product space. Let us define

$$\phi_{U \times V} = \phi_U \times \phi_V : U \times V \to \mathbb{R}^m \times \mathbb{R}^n = \mathbb{R}^{m+n}.$$

© The Author(s), under exclusive license to Springer Nature Singapore Pte Ltd. 2023
M. Majumdar and A. Bhattacharyya, *An Introduction to Smooth Manifolds*,
https://doi.org/10.1007/978-981-99-0565-2_4

If $(\tilde{U}, \phi_{\tilde{U}})$ and $(\tilde{V}, \phi_{\tilde{V}})$ are another pair of charts for M and N, respectively, then we can set the transition function

$$\phi_{\tilde{U} \times \tilde{V}} \circ \phi_{U \times V}^{-1} = (\phi_{\tilde{U}} \times \phi_{\tilde{V}}) \circ (\phi_U \times \phi_V)^{-1} = (\phi_{\tilde{U}} \circ \phi_U^{-1}) \times (\phi_{\tilde{V}} \circ \phi_V^{-1}).$$

Since each is a transition function from one of the two smooth atlases as already known on M and N, therefore, each of these factors is smooth. Since smoothness is determined componentwise, it follows that the product mapping is smooth as well. So we have an atlas making $M \times N$ a smooth manifold. It should also be clear that its dimension is $m + n$, as asserted.

Remark 4.3 Let G be a group, which is also a differentiable manifold of dimension n. It is easy to check that the map $(xy) \mapsto xy^{-1}$ is a smooth map from $2n$-dimensional manifold $G \times G$ to n-dimensional smooth manifold G. Thus, G is a Lie group.

Remark 4.4 Note that $\tau^2(x) = \tau(\tau(x)) = \tau(x^{-1}) = x$. Therefore, $\tau^2 = I$ is differentiable. Thus, G possesses $\tau^2 = I$ as diffeomorphism. Moreover, it possesses two other diffeomorphisms, viz. "Left Translation" and "Right Translation".

Example 4.1 Note that the n-dimensional space $\mathbb{R}^n$ is a differentiable manifold of n-dimension and is a group with respect to addition, defined as

$$x + y = (x^1 + y^1, \ldots, x^n + y^n)$$
$$-x = (-x^1, -x^2, \ldots, -x^n)$$
$$\text{where, } x = (x^1, x^2, \ldots, x^n), \quad y = (y^1, y^2, \ldots, y^n).$$

Furthermore, the operations $\phi(x + y) = x + y$, $\tau(x) = -x$ are C^∞-functions. Hence, $\mathbb{R}^n$ is a Lie group of dimension n.

Problem 4.1 $GL(n, \mathbb{R})$ *is a Lie group.*

Solution : *In reference to Problem 2.17, we have already proved that $GL(n, \mathbb{R})(\subset M(n, \mathbb{R}))$ forms a smooth manifold of dimension n^2. Note that $GL(n, \mathbb{R})$ forms a group under usual matrix multiplication. It only remains to prove that $GL(n, \mathbb{R})$ is a Lie group.*

For that, we define $\phi : GL(n, \mathbb{R}) \times GL(n, \mathbb{R}) \to GL(n, \mathbb{R})$ by $\phi(A, B) = AB$, so that

$$\phi(A, B) = AB = [\phi_j^i(AB)]_{i, j=1,2,\ldots,n},$$

where $\phi_j^i(AB)$ is the ij-th element of the matrix AB. Then, $\phi_j^i(AB) = \sum_{k=1}^n x_k^i(A) x_j^k(B)$, where $x_j^i(A)$ denotes the ij-th coordinate function on $GL(n, \mathbb{R})$. Thus, for all i, j; ϕ_j^i is a smooth function. Moreover, the inverse map $\tau : GL(n, \mathbb{R}) \to GL(n, \mathbb{R})$ defined by $\tau(A) = A^{-1}$ is also smooth. Hence $GL(n, \mathbb{R})$ is a Lie Group of dimension n^2.

For every $a \in G$, a mapping $L_a : G \to G$ defined by

$$L_a(x) = ax, \ \forall \ x \in G \tag{4.2}$$

is called a **left translation** on G. Similarly, a mapping $R_a : G \to G$ defined by

$$R_a(x) = xa, \ \forall \ x \in G \tag{4.3}$$

is called a **right translation on** G.

Exercises

Exercise 4.1 *Show that*

$$L_a L_b = L_{ab}, \quad R_a R_b = R_{ba}, \quad L_a R_b = R_b L_a. \tag{4.4}$$
$$L_a L_b \neq L_b L_a, \quad unless\ G\ is\ commutative. \tag{4.5}$$
$$L_{a^{-1}} = (L_a)^{-1}, \quad R_{a^{-1}} = (R_a)^{-1}. \tag{4.6}$$

It is to be noted that each L_a, R_a are C^∞ maps, as each L_a, $L_{a^{-1}}$, R_a, $R_{a^{-1}}$ is homeomorphism and differentiable from G onto G.

Examples

Example 4.2 The left translation $L_a : GL(n, \mathbb{R}) \to GL(n, \mathbb{R})$ is

$$x = (x^i_j) \to ax = (a^i_k x^k_j)$$

and the right translation $R_a : GL(n, \mathbb{R}) \to GL(n, \mathbb{R})$ is

$$x = (x^i_j) \to xa = (x^i_k a^k_j).$$

Example 4.3 The Lie group R^n is a commutative group. Hence, for every $a \in \mathbb{R}^n$, $L_a = R_a$. Also, the group operation is addition and the identity element is 0. So left translation is actually the parallel translation $x \mapsto x + a$ i.e. $L_a x = x + a = a + x$.

Problem 4.2 *Let $H = \left\{ \begin{pmatrix} 1 & x & y \\ 0 & 1 & z \\ 0 & 0 & 1 \end{pmatrix} : x, y, z \in \mathbb{R} \right\}$. Show that H admits a Lie group structure with usual matrix multiplication. Such H is called **Heisenberg Group**.*

Solution : *Let us define the map $\phi : H \to \mathbb{R}^3$ as $\begin{pmatrix} 1 & x & y \\ 0 & 1 & z \\ 0 & 0 & 1 \end{pmatrix} \mapsto (x, y, z)$. Note that the map ϕ is homeomorphic. Thus, $\{(H, \phi)\}$ forms an C^∞-atlas for H. If*

$$A = \begin{pmatrix} 1 & x & y \\ 0 & 1 & z \\ 0 & 0 & 1 \end{pmatrix}, \quad B = \begin{pmatrix} 1 & x' & y' \\ 0 & 1 & z' \\ 0 & 0 & 1 \end{pmatrix}, \quad \text{then } AB \in H.$$

Now $A^{-1} = \begin{pmatrix} 1 & -x & -y \\ 0 & 1 & -z \\ 0 & 0 & 1 \end{pmatrix} \in H, \quad \forall A \in H$. Again H is a group with respect to matrix multiplication. Moreover, let us define the maps

$$\Phi : H \times H \to H \quad \text{by } (A, B) \mapsto AB,$$

and the map

$$\psi : H \to H \quad \text{by } A \mapsto A^{-1}.$$

Then the map $\phi \circ \Phi \circ (\phi^{-1} \times \phi^{-1}) : \mathbb{R}^3 \times \mathbb{R}^3 \to \mathbb{R}^3$ defined by

$$(\phi \circ \Phi \circ (\phi^{-1} \times \phi^{-1}))((x, y, z), (a, b, c)) = (a + x, b + y, c + z)$$

and the map $\phi \circ \psi \circ \phi^{-1} : \mathbb{R}^3 \to \mathbb{R}^3$ given by

$$(\phi \circ \psi \circ \phi^{-1})(x, y, z) = (-x, -y, -z)$$

are C^∞. Thus H is a Lie group.

Problem 4.3 Let $\phi : G_1 \to G_2$ be a homomorphism of a Lie group G_1 to another Lie group G_2. Show that (i) $\phi \circ L_a = L_{\phi(a)} \circ \phi$ (ii) $\phi \circ R_b = R_{\phi(b)} \circ \phi$ for all $a, b \in G_1$.

Solution : (i) From the definition of group homomorphism $\phi : G_1 \to G_2$ given by

$$\phi(ab) = \phi(a)\phi(b), \quad \forall a, b \in G_1.$$

Now

$$(\phi \circ L_a)(x) = \phi(L_a(x)) = \phi(ax)$$
$$= \phi(a)\phi(x)$$
$$= (L_{\phi(a)} \circ \phi)(x).$$
$$\therefore \quad \phi \circ L_a = L_{\phi(a)} \circ \phi, \quad \forall x \in G_1.$$

(ii) Note that
$$(\phi \circ R_b)(x) = \phi(xb) = \phi(x)\phi(b) = (R_{\phi(b)} \circ \phi)(x).$$

Therefore, $\phi \circ R_b = R_{\phi(b)} \circ \phi, \quad \forall x \in G_1.$

Problem 4.4 Let ϕ be a non-identity $1 - 1$ map from G onto G. If $\phi \circ L_g = L_g \circ \phi$ holds for all $g \in G$, then there exists $h \in G$ such that $\phi = R_h$.

Solution : *As ϕ is a $1-1$ map, for every $a \in G$, there exists an unique element $b \in G$ such that $\phi(a) = b$. Again $e \in G$, ϕ is a $1-1$ map, there exists an unique element, say $h \in G$, such that $\phi(e) = h$, where $\phi(e) \neq e$, as ϕ is not an identity mapping. Now*

$$g = ge, \quad \forall \ g \in G$$
$$\therefore \quad \phi(g) = \phi(ge) = \phi(L_g(e)) = (\phi \circ L_g)(e) = (L_g \circ \phi)(e), \quad as \ given$$
$$= L_g(\phi(e))$$
$$= L_g(h) = gh = R_h g$$
$$\Rightarrow \quad \phi = R_h.$$

Exercises

Exercise 4.2 *Show that the set of all left(right) translations on a Lie group form a group.*

Exercise 4.3 *Let $G = \left\{ \begin{pmatrix} \alpha & 0 \\ \beta & 1 \end{pmatrix} : \alpha > 0, \beta \in \mathbb{R} \right\}$. Prove that G admits a Lie group structure with matrix multiplication.*

Exercise 4.4 *Let $G = \left\{ \begin{pmatrix} x & y \\ 0 & 1 \end{pmatrix} : x, y \in \mathbb{R}, x \neq 0 \right\}$. Prove that G admits a Lie group structure, with matrix multiplication.*

Exercise 4.5 *Let $G = \left\{ \begin{pmatrix} x_1 & 0 & x_2 \\ 0 & x_1 & x_3 \\ 0 & 0 & 1 \end{pmatrix} : x_1, x_2, x_3 \in \mathbb{R}, x_1 > 0 \right\}$. Prove that G admits a Lie group structure with respect to usual matrix multiplication.*

Exercise 4.6 *Let ψ be the diffeomorphism of G defined by $\psi(x) = x^{-1}$. Show that $\psi \circ L_g = R_{g^{-1}} \circ \psi$, $\psi \circ R_g = L_{g^{-1}} \circ \psi$.*

Exercise 4.7 *Prove that $\mathbb{R}$ is an abelian Lie group where the smooth maps $\phi : \mathbb{R} \times \mathbb{R} \to \mathbb{R}$ is defined by*

$$\phi(a, b) = a + b$$

and $\tau : \mathbb{R} \to \mathbb{R}$ is defined by

$$\tau(a) = -a.$$

4.2 Invariant Vector Field

We have already defined an invariant vector field under a transformation in §2.14. As each left translation and right translation on a Lie group G are transformations, we can similarly define the invariant vector fields on G.

A vector field X on a Lie group G is called a **left invariant vector field** on G if

$$(L_a)_* X_p = X_{L_{a(p)}} = X_{ap}, \ \forall \ p \in G, \tag{4.7}$$

where $(L_a)_*$ is the differential of the left translation L_a, for some fixed a in G.
    Using (2.35), we write

$$((L_a)_* X)_{L_a(p)} = X_{L_a(p)} \ i.e.$$
$$(L_a)_* X = X. \tag{4.8}$$

Similarly, for a right invariant vector field

$$(R_a)_* X = X. \tag{4.9}$$

Again, for every $f \in F(G)$, by virtue of (2.34), we have

$$\{(L_a)_* X_p\} f = X_p (f \circ L_a)$$
$$i.e. \ \left\{(L_a)_* X\right\}_{L_a(p)} f = X_p (f \circ L_a) \ \text{by (2.35)}.$$

If $L_a(p) = q$, then $p = a^{-1}q$ (refer to (4.6)). Thus

$$\{(L_a)_* X\}_q f = X_{a^{-1}q}(f \circ L_a). \tag{4.10}$$

Theorem 4.1 *A vector field X on a Lie group is left invariant if and only if*

$$(Xf) \circ L_a = X(f \circ L_a), \ \forall \ f \in F(G) \tag{4.11}$$

where for some fixed $a \in G$, L_a is the left translation of G.

Proof Let X be a left invariant vector field of a Lie group G. Then from (4.7), we
find

$$\{(L_a)_* X_p\} f = X_{L_a(p)} f, \ \forall \ f \in F(G)$$
$$= (Xf)L_a(p), \ \text{by (2.23)}$$
$$\text{or } X_p(f \circ L_a) = (Xf)L_a(p), \ \text{by (2.24)}$$
$$\text{or, } \{X(f \circ L_a)\}(p) = \big((Xf) \circ L_a\big)(p), \ \text{by (2.23)}.$$
$$\therefore \ X(f \circ L_a) = (Xf) \circ L_a, \ \forall \ p \in G.$$

The converse follows immediately.

Problem 4.5 *Find the left invariant vector fields on $\mathbb{R}^n$.*

Solution : *Let $p \in \mathbb{R}^n$ and $p = (x^1, x^2, \ldots, x^n)$. Any vector field X on the manifold $\mathbb{R}^n$ can be expressed uniquely as*

$$X = \sum \xi^i \frac{\partial}{\partial x^i}, \quad \text{where } \xi^i \in F(\mathbb{R}^n), \ i = 1, 2, 3, \ldots, n.$$

Furthermore

$$(Xf) \circ L_a = \left(\sum_i \xi^i \frac{\partial f}{\partial x^i} \right) \circ L_a, \quad a \in \mathbb{R}^n$$

$$\therefore \ \{(Xf) \circ L_a\}(p) = \left\{ \left(\sum_i \xi^i \frac{\partial f}{\partial x^i} \right) \circ L_a \right\}(p)$$

$$= \left(\sum_i \xi^i \frac{\partial f}{\partial x^i} \right) L_a(p)$$

$$= \left(\sum_i \xi^i \frac{\partial f}{\partial x^i} \right)(a + p), \quad \text{see Example 4.3}$$

$$\text{or } \ \{(Xf) \circ L_a\}(p) = \sum_i \xi^i (a + p) \frac{\partial f}{\partial x^i}(a + p).$$

Also

$$\{X(f \circ L_a)\}(p) = \left\{ \sum_i \xi^i \frac{\partial}{\partial x^i}(f \circ L_a) \right\}(p)$$

$$= \sum_i \xi^i(p) \frac{\partial}{\partial x^i}(f \circ L_a)(p)$$

$$= \sum_i \xi^i(p) \frac{\partial f}{\partial x^i}(L_a(p))$$

$$\text{or } \{X(f \circ L_a)\}(p) = \sum_i \xi^i(p) \frac{\partial f}{\partial x^i}(a + p).$$

In view of (4.11), we find

$$\sum_i \xi^i(a + p) \frac{\partial f}{\partial x^i}(a + p) = \sum_i \xi^i(p) \frac{\partial f}{\partial x^i}(a + p)$$

$$\text{or } \ \xi^i(a + p) = \xi^i(p), \quad i = 1, 2, 3, \ldots, n.$$

Thus, the functions ξ^i's are constants and hence all the left invariant vector fields on $\mathbb{R}^n$ are of the form

$$\xi^i \frac{\partial}{\partial x^i}, \quad \xi^i \in \mathbb{R}, \ i = 1, 2, 3, \ldots, n,$$

i.e.constant multiple of $\dfrac{\partial}{\partial x^i}$ *or the left invariant vector fields on* $\mathbb{R}^n$ *are constant vector fields.*

Problem 4.6 *If* X, Y *are left invariant vector fields on a Lie group* G, *so is* $[X, Y]$.

Solution : *From (4.8), we have* $(L_a)_* X = X$ *and* $(L_a)_* Y = Y$. *Now*

$$(L_a)_*[X, Y] = [(L_a)_* X, (L_a)_* Y] \ (see \ Exercise \ 2.30)$$
$$= [X, Y], \ from \ above.$$

Thus, $[X, Y]$ *is also left invariant vector field on* G.

ALTERNATIVE METHOD: *Taking into consideration (2.34), for every vector field* $[X, Y]$ *in* $\chi(G)$, *we have*

$$\{(L_a)_*[X, Y]\} f = [X, Y](f \circ L_a)$$
$$= X(Y(f \circ L_a)) - Y(X(f \circ L_a)), \ by \ (2.27)$$
$$= X\{((L_a)_* Y) f\} - Y\{((L_a)_* X) f\}$$
$$= X\{Yf\} - Y\{Xf\}, \ use \ (2.34)$$
$$= [X, Y] f$$
$$\therefore \ (L_a)_*[X, Y] = [X, Y], \ \forall \ f \in F(G).$$

We say $(G, [\ ,\ ])$ *is a Lie algebra over* $\mathbb{R}$ *if*

$$\begin{cases} (i) \ G \ is \ a \ vector \ space \ over \ \mathbb{R} \\ (ii) \ [,] : G \times G \to G \ is \ a \ bilinear \ map \\ (iii) \ [X, Y] = -[Y, X] : anti\text{-}commutative \\ (iv) \ [X, [Y, Z]] + [Y, [Z, X]] + [Z, [X, Y]] = \theta : Jacobi \ Identity. \end{cases} \quad (4.12)$$

Problem 4.7 *Show that the vector space* $\mathbb{R}^3$ *with the operation cross product of vectors is a Lie algebra.*

Solution : *Let* $\vec{x} = (x^1, x^2, x^3)$ *and* $\vec{y} = (y^1, y^2, y^3)$ *be any two vectors of* $\mathbb{R}^3$. *We define the cross product of* $\vec{x}$ *and* $\vec{y}$ *as follows:*

$$\vec{x} \times \vec{y} = (x^2 y^3 - x^3 y^2, x^3 y^1 - x^1 y^3, x^1 y^2 - x^2 y^1).$$

Then, for all $\lambda, \mu \in \mathbb{R}$, *it can be shown that*

$$\left. \begin{array}{l} (i) \ \ (\lambda \vec{x} + \mu \vec{y}) \times \vec{z} = \lambda \vec{x} \times \vec{z} + \mu \vec{y} \times \vec{z} \\ (ii) \ \ \vec{z} \times (\lambda \vec{x} + \mu \vec{y}) = \lambda \vec{z} \times \vec{x} + \mu \vec{z} \times \vec{y} \end{array} \right\} : Bilinearity$$

$$(iii) \ \ \vec{x} \times \vec{y} = -\vec{y} \times \vec{x}$$
$$(iv) \ \ (\vec{x} \times \vec{y}) \times \vec{z} + (\vec{y} \times \vec{z}) \times \vec{x} + (\vec{z} \times \vec{x}) \times \vec{y} = 0.$$

Thus, the real vector space $\mathbb{R}^3$ with the operation:Bilinearity of the cross product of vectors is a Lie algebra.

Remark 4.5 Note that the set of all C^∞-vector fields, denoted by $\chi(M)$, of the manifold M, forms a Lie algebra under the Lie bracket operation on vector fields.

Let $\mathbf{g}$ be the set of all left invariant vector fields on G. Then for every X, Y in $\mathbf{g}$,

$$(L_a)_*(cX + dY) = c(L_a)_*X + d(L_a)_*Y = cX + dY, \ \forall \, c, d \in \mathbb{R}.$$

Therefore, $cX + dY \in \mathbf{g}$. Hence, $\mathbf{g}$ is a linear space.
    Again, $(L_a)_*[X, Y] = [(L_a)_*X, (L_a)_*Y] = [X, Y]$ (refer to Exercise 2.30). Thus, $[X, Y] \in \mathbf{g}$ where $[X, Y] = -[Y, X]$. Further, it can be shown that

$$[X, [Y, Z]] + [Y, [Z, X]] + [Z, [X, Y]] = \theta.$$

Thus, the set of all left invariant vector fields, denoted by $\mathbf{g}$, on a Lie group G, is a Lie algebra.
    Now a Lie subalgebra h_1 of a Lie algebra h_2 is a vector subspace $h_1 \subset h_2$, that is closed under the bracket [,].
    Obviously, $\mathbf{g}$ is a **Lie subalgebra of the Lie algebra** $\chi(G)$ **of all** C^∞ **vector fields of the Lie group** G *i.e.*$\mathbf{g} \subset \chi(G)$

Remark 4.6 If $\mathbf{g}^*$ denotes the set of all right invariant vector fields on a Lie group G, it can be shown that $\mathbf{g}^*$ is also a Lie algebra.

The behaviour of a Lie group is determined largely by its behaviour in the neighbourhood of the identity element e. The behaviour can be represented by an algebraic structure on the tangent space at e.

Theorem 4.2 *As a vector space, the Lie subalgebra $\mathbf{g}$ of the Lie group G is isomorphic to the tangent space $T_e(G)$ at the identity element e of G i.e.$\mathbf{g} \cong T_e(G)$ (Fig. 4.1).*

Fig. 4.1 $\mathbf{g} \cong T_e(G)$

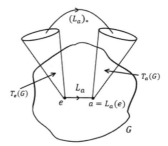

Proof Note that two vector spaces U and V are said to be isomorphic if a mapping $f : U \rightarrow V$ is linear and has an inverse f^{-1}. So, let us define a mapping $\phi : \mathbf{g} \rightarrow T_e(G)$ by

$$\phi(X) = X_e. \tag{4.13}$$

Clearly, ϕ is linear. Let us define $\phi^{-1} : T_e(G) \rightarrow \mathbf{g}$ by

$$\phi^{-1}(Y_e) = X. \tag{4.14}$$

Now for every $a \in G$, $L_a : G \rightarrow G$ is left translation and

$$(L_a)_* : T_e(G) \rightarrow T_{L_{(a)}(e)=a}(G)$$

is a differential mapping such that

$$(L_a)_* Y_e = X_a. \tag{4.15}$$

Now for any $s \in G$, we have

$$
\begin{aligned}
(L_s)_* X_{s^{-1}a} &= (L_s)_*(L_{s^{-1}a})_* Y_e, \;\; \text{by (4.15)} \\
&= (L_s \circ L_{s^{-1}a})_* Y_e, \;\; \text{by Problem 2.51} \\
&= (L_a)_* Y_e, \;\; \text{by (4.4)} \\
\text{or,} \;\; \big((L_s)_* X\big)_a &= X_a, \;\; \text{by (2.35), (4.15)} \\
\text{or} \;\; (L_s)_* X &= X, \;\; \forall \, a \in G.
\end{aligned}
$$

Therefore, $X \in \mathbf{g}$. Hence the mapping ϕ^{-1} is well-defined. Finally

$$
\begin{aligned}
(\phi \phi^{-1}) Y_e &= X_e, \;\; \text{by (4.13) \& (4.14)} \\
&= (L_e)_* Y_e, \;\; \text{by (4.15)} \\
&= Y_e, \;\; \text{where } (L_e)_* \text{ is the identity differential.} \\
(\phi^{-1}\phi)(X) &= \phi^{-1}\big((L_e)_* Y_e\big) = X, \;\; \text{by (4.13), (4.14) \& (4.15).}
\end{aligned}
$$

Thus, an inverse mapping ϕ^{-1} exists. Consequently,

$$\mathbf{g} \cong T_e(G).$$

Corollary 4.1 *If a Lie group G is of dimension n, then the dimension of Lie subalgebra $\mathbf{g}$ of the Lie group G is also n.*

Proof Left to the reader.

Problem 4.8 *If $\phi : \mathbf{g} \rightarrow T_e(G)$ is an isomorphism, $\mathbf{g}$ being the set of all left invariant vector fields of a Lie group G, then for $\tilde{X} = \phi(X) = X_e$, $X \in \mathbf{g}$, show that $[\tilde{X}, \tilde{Y}] = \widetilde{[X, Y]}$, $\forall \, Y \in \mathbf{g}$.*

Solution : *From the definition*

$$[\tilde{X}, \tilde{Y}] = \tilde{X}(\tilde{Y}) - \tilde{Y}(\tilde{X})$$
$$= X_e(Y_e) - Y_e(X_e), \quad \text{as defined}$$
$$= [X_e, Y_e].$$

Further

$$\widetilde{[X, Y]} = \{X(\widetilde{Y}) - Y(X)\}$$
$$= \widetilde{X(Y)} - \widetilde{Y(X)}$$
$$= X_e(Y_e) - Y_e(X_e)$$
$$= [X_e, Y_e].$$

Thus, $[\tilde{X}, \tilde{Y}] = \widetilde{[X, Y]}.$

Problem 4.9 *If* $C_{ij}^k(i, j, k = 1, 2, 3, \ldots, n)$ *are structure constants of a Lie group* G *with respect to the basis* $\{X_1, X_2, \ldots, X_n\}$ *of* **g**, *show that*

(i) $C_{ij}^k = -C_{ji}^k$, *where* $[X_i, X_j] = \sum C_{ij}^k X_k$, *each* $C_{ij}^k \in \mathbb{R}$.
(ii) $C_{ij}^k C_{ks}^t + C_{js}^k C_{ki}^t + C_{si}^k C_{kj}^t = 0$.

Solution : *It is given that* $C_{ij}^k(i, j, k = 1, 2, 3, \ldots, n)$ *are structure constants of a Lie group* G *with respect to the basis* $\{X_1, X_2, \ldots, X_n\}$ *of* **g**, *where* **g** *is the set of all left invariant vector fields on* G, *such that*

$$[X_i, X_j] = \sum_{k=1}^n C_{ij}^k X_k, \quad C_{ij}^k \in \mathbb{R}. \tag{4.16}$$

Note that $[X_i, X_j] = \theta$, *when* $i = j$. *So let* $i \neq j$. *As* $[X_i, X_j] = -[X_j, X_i]$, *by virtue of (4.16), we get*

$$\sum_{k=1}^n C_{ij}^k X_k = -\sum_{k=1}^n C_{ji}^k X_k$$
$$\text{or} \quad (C_{ij}^k + C_{ji}^k) = 0,$$

as $\{X_1, X_2, \ldots, X_n\}$ *is a basis and hence linearly independent. Thus,* $C_{ij}^k = -C_{ji}^k$. *Also, from the Jacobi identity, we have*

$$[[X_i, X_j], X_s] + [[X_j, X_s], X_i] + [[X_s, X_i], X_j] = \theta.$$

Using (4.16), we get

$$[\sum_{k=1}^{n} C_{ij}^{k} X_{k}, X_{s}] + [\sum_{k=1}^{n} C_{js}^{k} X_{k}, X_{i}] + [\sum_{k=1}^{n} C_{si}^{k} X_{k}, X_{j}] = \theta$$

$$or \ \sum_{k=1}^{n} C_{ij}^{k}[X_{k}, X_{s}] + \sum_{k=1}^{n} C_{js}^{k}[X_{k}, X_{i}] + \sum_{k=1}^{n} C_{si}^{k}[X_{k}, X_{j}] = \theta$$

$$\sum_{t}\sum_{k} C_{ij}^{k} C_{ks}^{t} X_{t} + \sum_{t}\sum_{k} C_{js}^{k} C_{ki}^{t} X_{t} + or \ \sum_{t}\sum_{k} C_{si}^{k} C_{kj}^{t} X_{t} = \theta, \quad by \ (4.16).$$

*As $\{X_{t} : t = 1, 2, 3, \ldots, n\}$ is a basis of **g** and hence linearly independent and thus*

$$C_{ij}^{k} C_{ks}^{t} + C_{js}^{k} C_{ki}^{t} + C_{si}^{k} C_{kj}^{t} = 0.$$

Problem 4.10 *Consider the product $T^{1} \times \mathbb{R}^{+}$ of the one-dimensional torus by the multiplicative group of positive numbers. Let $(\tilde{\alpha}, x_{1})$ denote the local coordinates. Prove that the vector field $X = \dfrac{\partial}{\partial\tilde{\alpha}} + x_{1}\dfrac{\partial}{\partial x_{1}}$ is left invariant.*

Solution : *For a fixed $(\tilde{\theta}, p) \in T^{1} \times \mathbb{R}^{+}$, the left translation of the product manifold, denoted by $L_{(\tilde{\theta},p)}$, is by definition*

$$L_{(\tilde{\theta},p)}(\tilde{\alpha}, x_{1}) = (\tilde{\theta} + \tilde{\alpha}, px_{1})$$

$$\therefore \ (L_{(\tilde{\theta},p)})_{*} = \begin{pmatrix} \frac{\partial(\tilde{\theta}+\tilde{\alpha})}{\partial\tilde{\alpha}} & \frac{\partial(\tilde{\theta}+\tilde{\alpha})}{\partial x_{1}} \\ \frac{\partial(px_{1})}{\partial\tilde{\alpha}} & \frac{\partial(px_{1})}{\partial x_{1}} \end{pmatrix} = \begin{pmatrix} 1 & 0 \\ 0 & p \end{pmatrix}$$

Given vector field $X \in T^{1} \times \mathbb{R}^{+}$ is left invariant if

$$(L_{(\tilde{\theta},p)})_{*} X_{(0,1)} = X_{L_{(\tilde{\theta},p)}(0,1)} = X_{(\tilde{\theta},p)}.$$

Now $X = \dfrac{\partial}{\partial\tilde{\alpha}} + x_{1}\dfrac{\partial}{\partial x_{1}}$, so $X_{(0,1)} = \dfrac{\partial}{\partial\tilde{\alpha}} + x_{1}\dfrac{\partial}{\partial x_{1}}\Big|_{(0,1)}$ and $X_{(\tilde{\theta},p)} = \dfrac{\partial}{\partial\tilde{\alpha}} + p\dfrac{\partial}{\partial x_{1}}\Big|_{(\tilde{\theta},p)}.$

Again

$$(L_{(\tilde{\theta},p)})_{*} X_{(0,1)} = \begin{pmatrix} 1 & 0 \\ 0 & p \end{pmatrix}\begin{pmatrix} 1 \\ 1 \end{pmatrix} = \begin{pmatrix} 1 \\ p \end{pmatrix} = \dfrac{\partial}{\partial\tilde{\alpha}} + p\dfrac{\partial}{\partial x_{1}}\Big|_{(\tilde{\theta},p)} = X_{(\tilde{\theta},p)}.$$

Thus, X is a left invariant vector field.

Problem 4.11 *Consider the Lie group G defined in Exercise 4.5. Show that*

(i) $\Big\{X = x_{1}\dfrac{\partial}{\partial x_{1}}, Y = x_{1}\dfrac{\partial}{\partial x_{2}}, Z = x_{1}\dfrac{\partial}{\partial x_{3}}\Big\}$ is a basis of left invariant vector fields of G.

(ii) *Find the structure constants, as defined by (4.16), of G with respect to the basis* $\{X, Y, Z\}$ *defined in (i) above.*

Solution : *(i) For a fixed* $(a_1, a_2, a_3) \in G$. *the left translation denoted by* $L_{(a_1,a_2,a_3)}$
is given by

$$L_{(a_1,a_2,a_3)}(x_1, x_2, x_3) = \begin{pmatrix} a_1 & 0 & a_2 \\ 0 & a_1 & a_3 \\ 0 & 0 & 1 \end{pmatrix} \begin{pmatrix} x_1 & 0 & x_2 \\ 0 & x_1 & x_3 \\ 0 & 0 & 1 \end{pmatrix} = \begin{pmatrix} a_1 x_1 & 0 & a_1 x_2 + a_2 \\ 0 & a_1 x_1 & a_1 x_3 + a_3 \\ 0 & 0 & 1 \end{pmatrix}$$

$$= (a_1 x_1, a_1 x_2 + a_2, a_1 x_3 + a_3)$$

$$\therefore \quad (L_{(a_1,a_2,a_3)})_* = \begin{pmatrix} a_1 & 0 & 0 \\ 0 & a_1 & 0 \\ 0 & 0 & a_1 \end{pmatrix}.$$

Let $e = (1, 0, 0)$ *denote the identity element of G. Then* $X = x_1 \dfrac{\partial}{\partial x_1}$ *is given by*

$$X_e = X_{(1,0,0)} = \frac{\partial}{\partial x_1}.$$

Similarly, $Y_e = \left(x_1 \dfrac{\partial}{\partial x_2}\right)_e = \left(\dfrac{\partial}{\partial x_2}\right)_e$ *and* $Z_e = \left(x_1 \dfrac{\partial}{\partial x_3}\right)_e = \left(\dfrac{\partial}{\partial x_3}\right)_e$. *Now*

$$X_{L_{(a_1,a_2,a_3)}}(1,0,0) = X_{(a_1,a_2,a_3)} = x_1 \frac{\partial}{\partial x_1}\bigg|_{(a_1,a_2,a_3)}.$$

Similarly

$$Y_{L_{(a_1,a_2,a_3)}}(1,0,0) = Y_{(a_1,a_2,a_3)} = x_1 \frac{\partial}{\partial x_2}\bigg|_{(a_1,b,a_3)} = a_1 \frac{\partial}{\partial x_2}, \quad and$$

$$Z_{L_{(a_1,a_2,a_3)}}(1,0,0) = Y_{(a_1,a_2,a_3)} = x_1 \frac{\partial}{\partial x_3}\bigg|_{(a_1,b,a_3)} = a_1 \frac{\partial}{\partial x_3}.$$

To show X, Y, Z *are left invariant vector fields, we have to show*

$$(L_{(a_1,a_2,a_3)})_* X_e = X_{L_{(a_1,a_2,a_3)}}(1,0,0) = X_{(a_1 \cdot 1, a_1 \cdot 0 + a_2, a_1 \cdot 0 + a_3)} = X_{(a_1,a_2,a_3)}.$$

Similarly

$$(L_{(a_1,a_2,a_3)})_* Y_e = Y_{(a_1,a_2,a_3)} \quad and \quad (L_{(a_1,a_2,a_3)})_* Z_e = Z_{(a_1,a_2,a_3)}.$$

Now

$$(L_{(a_1,a_2,a_3)})_* X_e = \begin{pmatrix} a_1 & 0 & 0 \\ 0 & a_1 & 0 \\ 0 & 0 & a_1 \end{pmatrix} \begin{pmatrix} 1 \\ 0 \\ 0 \end{pmatrix} = \begin{pmatrix} a_1 \\ 0 \\ 0 \end{pmatrix} = a_1 \frac{\partial}{\partial x_1} = X_{(a_1,a_2,a_3)}.$$

Thus, X is a left invariant vector field. Again

$$(L_{(a_1,a_2,a_3)})_* Y_e = \begin{pmatrix} a_1 & 0 & 0 \\ 0 & a_1 & 0 \\ 0 & 0 & a_1 \end{pmatrix} \begin{pmatrix} 0 \\ 1 \\ 0 \end{pmatrix} = \begin{pmatrix} 0 \\ a_1 \\ 0 \end{pmatrix} = a_1 \frac{\partial}{\partial x_2} = Y_{(a_1,a_2,a_3)},$$

$$(L_{(a_1,a_2,a_3)})_* Z_e = \begin{pmatrix} a_1 & 0 & 0 \\ 0 & a_1 & 0 \\ 0 & 0 & a_1 \end{pmatrix} \begin{pmatrix} 0 \\ 0 \\ 1 \end{pmatrix} = \begin{pmatrix} 0 \\ 0 \\ a_1 \end{pmatrix} = a_1 \frac{\partial}{\partial x_3} = Z_{(a_1,a_2,a_3)}.$$

Thus, Y and Z are left invariant vector fields and $\{X, Y, Z\}$ are linearly independent at e. Thus, $\{X, Y, Z\}$ is a basis of $\mathbf{g}$, where $\mathbf{g}$ is a left invariant vector fields of G.

(ii) *If we denote $X = x_1 \dfrac{\partial}{\partial x_1}$ by X_1, $Y = x_1 \dfrac{\partial}{\partial x_2}$ by X_2 and $Z = x_1 \dfrac{\partial}{\partial x_3}$ by X_3, then by (4.13), we see that*

$$[X_i, X_j] = \sum_{k=1}^{3} C_{ij}^k X_k, \quad i \neq j.$$

Now

$$[X_1, X_2] = [x_1 \frac{\partial}{\partial x_1}, x_1 \frac{\partial}{\partial x_2}] = x_1 \frac{\partial}{\partial x_1}\left(x_1 \frac{\partial}{\partial x_2}\right) - x_1 \frac{\partial}{\partial x_2}\left(x_1 \frac{\partial}{\partial x_1}\right)$$

$$= x_1 \frac{\partial}{\partial x_2} + x_1^2 \frac{\partial^2}{\partial x_1 \partial x_2} - x_1^2 \frac{\partial^2}{\partial x_1 \partial x_2}$$

$$= x_1 \frac{\partial}{\partial x_2}.$$

$$\therefore \quad C_{12}^2 X_2 = X_2, \quad \text{by above}$$

i.e. $C_{12}^2 = 1$, as $X_2 \equiv Y$ is linearly independent. By Problem 4.9(i), $C_{21}^2 = -C_{12}^2 = -1$. Similarly, it can be shown that $[X_1, X_3] = X_3$, i.e. $C_{13}^3 = 1$ and $C_{31}^3 = -1$ and $[X_2, X_3] = 0$. Thus, with respect to the basis $\{X, Y, Z\}$, the structure given by

$$C_{12}^2 = -C_{21}^2 = C_{13}^3 = -C_{31}^3 = 1.$$

Exercises

Exercise 4.8 *Prove the converse of Theorem 4.1.*

Exercise 4.9 *Find the left invariant vector fields on* $\mathbb{R}$.

Exercise 4.10 *If e is the identity element of a Lie group G and* $T_e(G)$ *is the tangent space to G at e, show that*

$$(L_a)_* X_e = X_a,$$

where X is a left invariant vector field.

Exercise 4.11 *Consider the Problem 4.2. Show that*

(i) $\{X = \dfrac{\partial}{\partial x}, Y = \dfrac{\partial}{\partial y}, Z = x\dfrac{\partial}{\partial y} + \dfrac{\partial}{\partial z}\}$ *is a basis of left invariant fields of H.*

(ii) *Find the structure constants of H with respect to the basis* $\{X, Y, Z\}$.

Answer

4.11. $C_{13}^2 = -C_{31}^2 = 1.$

4.3   Invariant Differential Form

A differential form ω on a Lie group G is said to be **left invariant** if

$$\left(L_a^* \omega_{L_a(p)}\right) = \omega_p, \ \forall\, p \in G \tag{4.17}$$

where L_a^* is the pull-back of L_a defined in §3.4.
   We write it as

$$L_a^* \omega = \omega. \tag{4.18}$$

Similarly, a differential form ω on a Lie group G is said to be **right invariant** if

$$R_a^* \omega = \omega. \tag{4.19}$$

A differential form, which is both left and right invariant is said to be **bi-invariant differential form**.

Problem 4.12 *Show that if ω is a left invariant form, then $d\omega$ is also so.*

Solution : *From Theorem 3.9, we see that*

$$d(L_a^* \omega) = L_a^*(d\omega), \ \forall\, \omega$$
$$or, \ d\omega = L_a^*(d\omega), \ by\ (4.18).$$

Hence, $d\omega$ is also left invariant by (4.18).

Problem 4.13 *Prove that a 1-form ω on a Lie group G is left invariant if and only if for every left invariant vector field X on G, $\omega(X)$ is a constant function on G.*

Solution : *For every given 1-form ω on G, $L_a^*\omega$ will be pull-back 1-form. Hence, by (3.27), we find*

$$\{L_a^*(\omega_{L_a(p)})\}(X_p) = \omega_{L_a(p)}((L_a)_*X_p), \ \forall\, p \in G, \ \forall\, X \in \mathbf{g},$$

$\mathbf{g}$ being the set of all invariant vector fields, $(L_a)_*X_p$ being the differential of X. Consequently, by (4.7), we write

$$\{L_a^*(\omega_{L_a(p)})\}(X_p) = \omega_{L_a(p)}(X_{L_a(p)}). \tag{4.20}$$

Let us now consider ω to be left invariant. Then in view of (4.17), one gets from above

$$\omega_p(X_p) = \omega_{ap}(X_{ap}).$$

Taking $p = e \in G$, one gets the desired result, i.e. $\omega(X)$ is a constant function.

For the converse part, let $\omega(X)$ be a constant function on G. Then for fixed $a \in G$ and arbitrary $p \in G$, one must have

$$\begin{aligned}
\omega_p(X_p) &= \omega_{ap}(X_{ap}) \\
&= \omega_{L_a(p)}(X_{L_a(p)}) \\
&= \{L_a^*(\omega_{L_a(p)})\}(X_p), \ \ see \ (4.20) \\
\therefore \ L_a^*(\omega_{L_a(p)}) &= \omega_p, \ \forall\, X_p \in \mathbf{g}.
\end{aligned}$$

Thus, by (4.17), the 1-form ω is left invariant.

Problem 4.14 *Prove that the set of all left invariant forms on a Lie group G forms an algebra over $\mathbb{R}$.*

Solution : *Let A be the set of all left invariant forms on a Lie group G. We wish to show that:*

(i) A is a linear space over $\mathbb{R}$.
(ii) the mapping $A \times A \to A$, defined as $(\omega, \mu) \mapsto \omega \wedge \mu$ is bilinear.
(iii) the operation '$\wedge$' is skew-symmetric.
(iv) $\begin{cases} \omega \wedge (\mu + \gamma) = \omega \wedge \mu + \omega \wedge \gamma \\ (\omega + \mu) \wedge \gamma = \omega \wedge \gamma + \mu \wedge \gamma. \end{cases}$

Note that

$$\begin{aligned}
L_a^*(c\omega + d\mu) &= c\, L_a^*\omega + d\, L_a^*\mu, \ \ as \ L_a^* \ is \ linear, \ c, d \in \mathbb{R} \\
&= c\omega + d\mu, \ where \ \omega, \mu \ are \ left \ invariant.
\end{aligned}$$

Thus, A is a linear space over $\mathbb{R}$.

Further, it can be shown that

$$b\omega \wedge \mu = b(\omega \wedge \mu) = \omega \wedge b\mu, \quad \forall b \in \mathbb{R}.$$

Thus, the mapping is bilinear. Also, $\omega \wedge \mu = -\mu \wedge \omega$. Moreover, the set A satisfies (iv). Thus, A is an algebra over $\mathbb{R}$.

Problem 4.15 *Let $L_a : S^1 \to S^1$ be given by*

$$L_a(x, y) = \{(\cos t)x - (\sin t)y, (\sin t)x + (\cos t)y\}$$

where $a = (\cos t, \sin t) \in S^1 \subset \mathbb{R}^2$. If $\omega = -ydx + xdy$ on S^1, show that $(L_a)^*\omega = \omega$,
S^1 being the unit circle.

Solution : *Here*

$$\begin{aligned}
L_a^*\omega &= L_a^*(-ydx + xdy) \\
&= L_a^*(-ydx) + L_a^*(xdy) \\
&= -\{(\sin t)x + (\cos t)y\}d\{(\cos t)x - (\sin t)y\} \\
&\quad + \{(\cos t)x - (\sin t)y\}d\{(\sin t)x + (\cos t)y\} \\
&= -\{(\sin t)x + (\cos t)y\}\{(\cos t)dx - (\sin t)dy\} \\
&\quad + \{(\cos t)x - (\sin t)y\}\{(\sin t)dx + (\cos t)dy\} \\
&= xdy - ydx \\
&= \omega.
\end{aligned}$$

Thus, ω is left invariant 1-form on S^1.

Theorem 4.3 *If g is Lie subalgebra of a Lie group G and g^* denotes the set of all left invariant forms on G, then*

$$d\omega(X, Y) = -\frac{1}{2}\omega([X, Y]), \quad \forall X, Y \in g, \omega \in g^*.$$

Proof From Theorem 3.4, we see that if ω is a 1-form, then

$$d\omega(X, Y) = \frac{1}{2}\{X(\omega(Y)) - Y(\omega(X)) - \omega([X, Y])\}, \quad \forall X, Y \in \chi(G).$$

Now if $X, Y \in g$, $\omega \in g^*$, then by Problem 4.13, $\omega(X), \omega(Y)$ are constant functions on G. Taking help of Exercise 2.30, we see that

$$X(\omega(Y)) = 0 = Y(\omega(X)).$$

Thus $d\omega(X, Y) = -\frac{1}{2}\omega([X, Y]).$

Remark 4.7 Such an equation is called **Maurer–Cartan Equation**.

Problem 4.16 *Show that* $d\omega^i = -\dfrac{1}{2}\displaystyle\sum_{j,k} C^i_{jk}\,\omega^j \wedge \omega^k$, *where* C^i_{jk}*'s are defined in Exercise 4.9(i).*

Solution : *If* $\{X_1, X_2, \ldots, X_n\}$ *is a basis of* **g** *and* $\{\omega^1, \omega^2, \ldots, \omega^n\}$ *is the dual basis of* **g***, *then*

$$\omega^i(X_j) = \delta^i_j. \tag{4.21}$$

Hence, by virtue of the last theorem, we obtain

$$\begin{aligned}
d\omega^i(X_j, X_k) &= -\frac{1}{2}\omega^i([X_j, X_k]) \\
&= -\frac{1}{2}\omega^i\{\sum C^m_{jk}X_m\}, \quad see \ (4.16) \\
&= -\frac{1}{2}\sum_m C^m_{jk}\omega^i(X_m), \quad as \ C^m_{jk} \in \mathbb{R} \\
d\omega^i(X_j, X_k) &= -\frac{1}{2}C^i_{jk}.
\end{aligned} \tag{4.22}$$

Now

$$\begin{aligned}
\sum_{m,n} C^i_{mn}(\omega^m \wedge \omega^n)(X_j, X_k) &= \frac{1}{2}\sum_{m,n} C^i_{mn}\{\omega^m(X_j)\omega^n(X_k) - \omega^m(X_k)\omega^n(X_j)\}, \quad by \ (3.17) \\
&= \frac{1}{2}\sum_{m,n} C^i_{mn}(\delta^m_j\delta^n_k - \delta^m_k\delta^n_j), \quad by \ (4.21) \\
&= \frac{1}{2}(C^i_{jk} - C^i_{kj}) \\
&= C^i_{jk}, \quad see \ Problem \ 4.9(i)
\end{aligned}$$

Thus, (4.22) reduces to

$$d\omega^i(X_j, X_k) = -\frac{1}{2}\sum_{m,n} C^i_{mn}(\omega^m \wedge \omega^n)(X_j, X_k)$$

$$or \ d\omega^i = -\frac{1}{2}\sum_{m,n} C^i_{mn}(\omega^m \wedge \omega^n), \quad \forall \ X_j, X_k$$

$$i.e. \ d\omega^i = -\frac{1}{2}\sum_{j,k} C^i_{jk}(\omega^j \wedge \omega^k).$$

Problem 4.17 *Prove that* $d\omega^i = \sum\limits_{\substack{j,k \\ j<k}}^{i} C_{jk}^i \omega^k \wedge \omega^j.$

Solution : *Let us consider* $j, k = 1, 2, 3$. *Then*

$$\sum_{j,k=1,2,3} C_{jk}^i \, \omega^j \wedge \omega^k = C_{12}^i \omega^1 \wedge \omega^2 + C_{13}^i \omega^1 \wedge \omega^3 + C_{21}^i \omega^2 \wedge \omega^1$$

$$+ C_{23}^i \omega^2 \wedge \omega^3 + C_{31}^i \omega^3 \wedge \omega^1 + C_{32}^i \omega^3 \wedge \omega^2$$

$$= 2C_{12}^i \omega^1 \wedge \omega^2 + 2C_{13}^i \omega^1 \wedge \omega^3 + 2C_{23}^i \omega^2 \wedge \omega^3,$$

as $\omega^i \wedge \omega^i = 0, \omega^i \wedge \omega^j = -\omega^j \wedge \omega^i, C_{jk}^i = -C_{kj}^i$. *Thus*

$$\sum_{j,k=1,2,3} C_{jk}^i \omega^j \wedge \omega^k = 2 \sum_{\substack{j,k \\ j<k}}^{i} C_{jk}^i \, \omega^j \wedge \omega^k.$$

Hence, from Problem 4.16, we have

$$d\omega^i = -\frac{1}{2} \times 2 \sum_{\substack{j,k \\ j<k}}^{i} C_{jk}^i \, \omega^j \wedge \omega^k$$

or $d\omega^i = \sum\limits_{\substack{j,k \\ j<k}}^{i} C_{jk}^i \, \omega^k \wedge \omega^j.$

Problem 4.18 *Consider* $G = \left\{ \begin{pmatrix} x_1 & x_2 \\ 0 & 1 \end{pmatrix} \middle| x_1, x_2 \in \mathbb{R}, x_1 \neq 0 \right\}$ *which is a Lie subgroup of* $GL(2, \mathbb{R})$.

(i) *Show that* $\omega = a^{-1}da$ *is a left invariant 1-form, where* $a = \begin{pmatrix} x_1 & x_2 \\ 0 & 1 \end{pmatrix}$.

(ii) *Show that* $\{\omega^1 = \dfrac{dx_1}{x_1}, \omega^2 = \dfrac{dx_2}{x_1}\}$ *is a basis of left invariant 1-form* g^* *and find the structure constants of* G *with respect to* $\{\omega^1, \omega^2\}$.

(iii) *Show that* $d\omega + \omega \wedge \omega = 0$.

Solution : (i) *Here* $a = \begin{pmatrix} x_1 & x_2 \\ 0 & 1 \end{pmatrix}$, $|a| = x_1$, $adj\, a = \begin{pmatrix} 1 & -x_2 \\ 0 & x_1 \end{pmatrix}$, $a^{-1} = \dfrac{1}{x_1} \begin{pmatrix} 1 & -x_2 \\ 0 & x_1 \end{pmatrix}$.

Also, $da = \begin{pmatrix} dx_1 & dx_2 \\ 0 & 0 \end{pmatrix}$. Thus, $\omega = a^{-1}da = \dfrac{1}{x_1}\begin{pmatrix} 1 & -x_2 \\ 0 & x_1 \end{pmatrix}\begin{pmatrix} dx_1 & dx_2 \\ 0 & 0 \end{pmatrix}$

$= \dfrac{1}{x_1}\begin{pmatrix} dx_1 & dx_2 \\ 0 & 0 \end{pmatrix}$.

Let us choose $e = \begin{pmatrix} 1 & 0 \\ 0 & 1 \end{pmatrix}$ and arbitrary $q = \begin{pmatrix} r & s \\ 0 & 1 \end{pmatrix}$. Thus,

$$L_q a = \begin{pmatrix} r & s \\ 0 & 1 \end{pmatrix}\begin{pmatrix} x_1 & x_2 \\ 0 & 1 \end{pmatrix} = \begin{pmatrix} rx_1 & rx_2 + s \\ 0 & 1 \end{pmatrix}, \quad (L_q)_* = \begin{pmatrix} r & 0 \\ 0 & r \end{pmatrix}.$$

Now, $\omega_p = \dfrac{1}{p}\begin{pmatrix} dx_1 & dx_2 \\ 0 & 0 \end{pmatrix}$, $\omega_e = \begin{pmatrix} dx_1 & dx_2 \\ 0 & 0 \end{pmatrix}$. We are to show $(L_p)^*\omega_{L_{pe}} = \omega_e$,

i.e. $(L_p)^*\omega_p = \omega_e$. Now

$$(L_p)^*\omega_p = \left\{\begin{pmatrix} p & 0 \\ 0 & p \end{pmatrix}\begin{pmatrix} \frac{1}{p} \\ 0 \end{pmatrix}, \begin{pmatrix} p & 0 \\ 0 & p \end{pmatrix}\begin{pmatrix} 0 \\ \frac{1}{p} \end{pmatrix}\right\} = \left\{\begin{pmatrix} 1 \\ 0 \end{pmatrix}, \begin{pmatrix} 0 \\ 1 \end{pmatrix}\right\} = \begin{pmatrix} dx_1 & dx_2 \\ 0 & 0 \end{pmatrix} = \omega_e.$$

Thus, $\omega = a^{-1}da$ is a left invariant 1-form.

(ii) Note that

$$d\omega^1 = d\left(\frac{dx_1}{x_1}\right) = d\left(\frac{1}{x_1}\right)\wedge dx_1 = 0$$

$$d\omega^2 = d\left(\frac{dx_2}{x_1}\right) = d\left(\frac{1}{x_1}\right)\wedge dx_2 - -\frac{1}{x_1^2}dx_1\wedge dx_2$$

$$= -\omega^1\wedge\omega^2.$$

Taking advantage of Maurer–Cartan Equation, we obtain

$$d\omega^i = -\sum_{j,k} C^i_{jk}\omega^j\wedge\omega^k,$$

for $i = 2$, $d\omega^2 = -C^2_{12}\omega^1\wedge\omega^2$. Comparing, we find that $C^2_{12} = 1 = -C^2_{21}$.

(iii) Here $\omega = \dfrac{1}{x_1}\begin{pmatrix} dx_1 & dx_2 \\ 0 & 0 \end{pmatrix}$. Therefore

$$\omega\wedge\omega = \dfrac{1}{x_1}\begin{pmatrix} dx_1 & dx_2 \\ 0 & 0 \end{pmatrix}\wedge\dfrac{1}{x_1}\begin{pmatrix} dx_1 & dx_2 \\ 0 & 0 \end{pmatrix} = \dfrac{1}{x_1^2}\begin{pmatrix} 0 & dx_1\wedge dx_2 \\ 0 & 0 \end{pmatrix}.$$

Further

$$dw = d\left(\frac{1}{x_1}\right) \wedge \begin{pmatrix} dx_1 & dx_2 \\ 0 & 0 \end{pmatrix}$$

$$= -\frac{1}{x_1^2} dx_1 \wedge \begin{pmatrix} dx_1 & dx_2 \\ 0 & 0 \end{pmatrix}$$

$$= -\frac{1}{x_1^2} \begin{pmatrix} 0 & dx_1 \wedge dx_2 \\ 0 & 0 \end{pmatrix}$$

$$= -\omega \wedge \omega, \text{ from above.}$$

Thus, $d\omega + \omega \wedge \omega = 0$.

Exercises

Exercise 4.12 *If ω_1, ω_2 are left invariant differential forms, prove that $\omega_1 \wedge \omega_2$ is also so.*

Exercise 4.13 *Prove that a r-form ω on a Lie group G is left invariant if and only if for every left invariant vector fields X_i's$(1 \leq i \leq r)$ on G, $\omega(X_1, X_2, \ldots, X_r)$ is a constant function on G.*

Exercise 4.14 *Let $\phi : G \to G$ be such that $\phi(a) = a^{-1}$, $\forall a \in G$. Show that a form ω is left invariant if and only if $\phi^*\omega$ is right invariant.*

Exercise 4.15 *If g^* denotes the dual space of g, prove that $A \cong g^*$, where the set A is defined in the solution of Problem 4.14.*

4.4 Automorphism

Let G_1 and G_2 be Lie group. A map $f : G_1 \to G_2$ is said to be a **Lie group homomorphism** if f is a C^∞ map and for all h, x in G_1,

$$f(hx) = f(h)f(x). \tag{4.23}$$

We can also write it as

$$f \circ L_h = L_{f(h)} \circ f, \ \forall x \in G_1. \tag{4.24}$$

Let e_{G_1} and e_{G_2} be the identity elements of G_1 and G_2, respectively. Taking h, x in (4.23) to be the identity e_{G_1}, it follows that

$$f(e_{G_1}) = e_{G_2}. \tag{4.25}$$

So a group homomorphism always maps the identity to identity.

If moreover, f is bijective, then f is said to be a **Lie group isomorphism**.

For every $a \in G$, a mapping $\sigma_a : G \to G$, defined by

$$\sigma_a(x) = axa^{-1} \tag{4.26}$$

is said to be an **inner automorphism** if

$$\begin{cases} (i) \ \ \sigma_a \text{ is bijective} \\ (ii) \ \sigma_a(xy) = \sigma_a(x)\sigma_a(y) \end{cases} \tag{4.27}$$

Such σ_a is **written as** ada.

An inner automorphism of a Lie group G is defined by

$$(ada)(x) = axa^{-1}, \ \ \forall x \in G. \tag{4.28}$$

Problem 4.19 *If the Lie group G is defined by $G = \left\{ \begin{pmatrix} x & y \\ 0 & x \end{pmatrix} : x, y \in \mathbb{R}, x > 0 \right\}$, verify whether the map $f : G \to \mathbb{R}^3$ defined by $f \begin{pmatrix} x & y \\ 0 & x \end{pmatrix} = (x, y, x - y)$ is a Lie group homomorphism or not.*

Solution : *For $\begin{pmatrix} x' & y' \\ 0 & x' \end{pmatrix} \in G$, we have $\begin{pmatrix} x & y \\ 0 & x \end{pmatrix} \begin{pmatrix} x' & y' \\ 0 & x' \end{pmatrix} = \begin{pmatrix} xx' & xy' + yx' \\ 0 & xx' \end{pmatrix}$. From the hypothesis,*

$$f \begin{pmatrix} x & y \\ 0 & x \end{pmatrix} \begin{pmatrix} x' & y' \\ 0 & x' \end{pmatrix} = (xx', xy' + yx', xx' - xy' - yx').$$

To show that f is a Lie group homomorphism, our claim is

$$f(AB) = f(A) + f(B), \quad A, B \in G, \ \text{where } G \text{ is a lie group.}$$

Now

$$f \begin{pmatrix} x & y \\ 0 & x \end{pmatrix} + f \begin{pmatrix} x' & y' \\ 0 & x' \end{pmatrix} \neq f \begin{pmatrix} x & y \\ 0 & x \end{pmatrix} \begin{pmatrix} x' & y' \\ 0 & x' \end{pmatrix}.$$

Thus, f is not a Lie group homomorphism.

Exercises

Exercise 4.16 *Show that if G is a Lie group, then the map $I_h : G \to G$ for every $h \in G$ defined by $I_h(x) = hxh^{-1}$, $x \in G$ is an automorphism.*

Exercise 4.17 *Show that*

$$ada = L_a R_{a^{-1}} = R_{a^{-1}} L_a. \tag{4.29}$$

Exercise 4.18 *Let $H = \left\{ \begin{pmatrix} 1 & x & y \\ 0 & 1 & z \\ 0 & 0 & 1 \end{pmatrix} : x, y, z \in \mathbb{R} \right\}$ be the Lie group and the map*
$f : H \rightarrow \mathbb{R}$ *defined by $A \mapsto f(A) = x + y + z$. Is it a Lie group homomorphism?*

Answer

4.18. No.

Let $\mathbf{g}_1$ and $\mathbf{g}_2$ be Lie subalgebra of the Lie group G. A mapping $f : \mathbf{g}_1 \rightarrow \mathbf{g}_2$ is said to be a **Lie algebra homomorphism** if f is a linear mapping and

$$f[X, Y] = [fX, fY]. \tag{4.30}$$

Moreover, if f is bijective then f is said to be **Lie algebra isomorphism**.

Problem 4.20 *Let $G_1 = GL(n, \mathbb{R})$ and $G_2 = GL(1, \mathbb{R})$. Then the map f given by $f(A) = \det A$, $A \in G_1$ is a homomorphism.*

Solution : *Clearly*

$$f(cA + B) = cf(A) + f(B), \ \forall c \in \mathbb{R}, \ A, B \in G_1,$$

which implies f is linear. Furthermore, $f(AB) = \det(AB) = \det A \det B = f(A) f(B)$ implies f is a homomorphism.

Problem 4.21 *If $f : G_1 \rightarrow G_2$ is a Lie group homomorphism and X is a left invariant vector field on G_1, prove that the left invariant vector field $f_* X$ on G_2 is f-related to the left invariant vector field X.*

Solution : *For $h \in G_1$, $f(h) \in G_2$. For every $X_h \in \mathbf{g}_1$, $f_*(X_h) \in \mathbf{g}_2$. To show that $f_* X$ is f-related to X, we need to show $f_*(X_h) = (f_* X)_{f(h)}$, by (2.38). Note that*

$$(L_h)_* X_e = X_{L_h(e)} = X_h, \quad X \in \mathbf{g}_1, \quad \mathbf{g}_1 \cong T_{e_1}(G_1).$$

Thus

$$
\begin{aligned}
f_*(X_h) &= f_*((L_h)_* X_e) = (f \circ L_h)_* X_e, \\
&= (L_{f(h)} \circ f)_*(X_e), \quad by \ (4.24) \\
&= (L_{f(h)})_* \{f_*(X_e)\} \\
&= \{f_*(X)\}_{L_{f(h)}(e)}, \quad by \ (4.7).
\end{aligned}
$$

$$\therefore \quad f_*(X_h) = \{f_*(X)\}_{f(h)}.$$

Hence, the left invariant vector field $f_* X$ on G_2 is f-related to the left invariant vector field X on G_1.

Theorem 4.4 *Let $f : G_1 \to G_2$ be a Lie group homomorphism. Then the induced map $f_* : T_e(G_1) \to T_{e'}(G_2)$ is a homomorphism between the Lie algebras of the Lie group, where e, e' are respectively the identity elements of G_1 and G_2.*

Proof For the identity element $e \in G_1$,

$$f(e) = e',$$

e' being the identity element of G_2. In view of Theorem 4.2, we can write

$$\mathbf{g}_1 \cong T_e(G_1), \quad \text{and} \quad \mathbf{g}_2 \cong T_{e'}(G_2),$$

where $\mathbf{g}_1, \mathbf{g}_2$ are respectively the Lie algebras of G_1 and G_2.
Now, let $X_e \in T_e(G_1)$ be such that

$$f_*(X_e) = Y_{e'} \in T_{e'}(G_2).$$

Then for any $a \in G_1$, $(L_a)_* X_e = X_{L_a(e)} = X_a$. Therefore

$$
\begin{aligned}
f_*(X_a) &= f_*\{(L_a)_* X_e\} \\
&= (f \circ L_a)_* X_e, \text{ as } (fg)_* = f_* \circ g_* \\
&= (L_{f(a)} \circ f)_* X_e, \text{ by (4.24)} \\
&= (L_{f(a)})_* Y_{e'} \\
&= Y_{f(a)}, \text{ refer to the definition of left invariant vector field}
\end{aligned}
$$

Thus, the image of a left invariant vector field on G_1 under f_* is a left invariant vector field on G_2.
Again, we know that

$$
\begin{aligned}
f_*[X_1, X_2] &= [f_* X_1, f_* X_2] \\
&= [Y_1, Y_2],
\end{aligned}
$$

where $f_* X_i = Y_i, i = 1, 2$. Thus, f_* is a homomorphism between the Lie algebras of two spaces and this completes the proof.

Theorem 4.5 *Every inner automorphism of a Lie group G induces an automorphism of the Lie algebra $\mathbf{g}$ of G.*

Proof For every $a \in G$, let us denote the inner automorphism on G by

$$(ada)(x) = axa^{-1}, \quad \forall x \in G.$$

Now every $ada : G \to G$ induces a differential mapping

$$(ada)_* : T_e(G) \to T_{ada(e)}(G) \equiv T_e(G), \text{ see (4.28)}.$$

Such a mapping is a linear mapping, and from Theorem 4.2, we have $\mathbf{g} \cong T_e(G)$. Thus, to prove the theorem, we need to prove the following:

(i) $(ada)_* : \mathbf{g} \to \mathbf{g}$ is a well-defined mapping.
(ii) $(ada)_*$ is bijective.
(iii) $(ada)_*$ is homomorphism.

By (4.29), we get

$$(ada)_* = (R_{a^{-1}})_*, \quad \forall Y \in \mathbf{g}. \tag{4.31}$$

Now

$$
\begin{aligned}
(L_p)_*\{(R_{a^{-1}})_* Y\} &= (L_p \circ R_{a^{-1}})_* Y, \quad p \in G \\
&= (R_{a^{-1}})_* Y, \quad \text{see (4.29)}.
\end{aligned}
$$

Thus, $(R_{a^{-1}})_* Y \in \mathbf{g}$. Consequently, $(ada)_*$ is a well-defined mapping that proves (i). Also, $(ada)_*$ is a linear mapping and by Exercise 2.30, the (iii) follows immediately.

Finally, let $(ada)_* X = (ada)_* Y$. Then from (4.31), it follows that $X = Y$ and hence $(ada)_*$ is injective. For surjectivity, let us set $(ada^{-1})_* X = Y$, as for every $a, a^{-1} \in G$. Now

$$
\begin{aligned}
(L_s)_* Y &= (L_s)_* ((ada^{-1})_* X) \\
&= (L_s)_* \circ ((L_{a^{-1}} \circ R_a)_* X), \quad \text{by (4.29)} \\
&= (L_s)_* \circ ((R_a)_* X), \quad \text{by (4.4)} \\
&= (R_a)_* X, \quad \text{by (4.4) and Problem 2.51} \\
&= Y, \quad \text{by (4.31) and as assumed}.
\end{aligned}
$$

Thus, $Y \in \mathbf{g}$. Also

$$
\begin{aligned}
(ada)_* Y &= (L_a \circ R_{a^{-1}})_* Y, \quad \text{by (4.29)} \\
&= (L_a \circ R_{a^{-1}})_* (ada^{-1})_* X, \quad \text{as set} \\
&= (L_a \circ R_{a^{-1}})_* (R_a \circ L_a^{-1})_* X, \quad \text{by (4.29)} \\
&= X, \quad \text{by Problem 2.51}
\end{aligned}
$$

which proves $(ada)_*$ is surjective and consequently, $(ada)_*$ is a bijective mapping.

Thus, the induced map $(ada)_* : \mathbf{g} \to \mathbf{g}$ is a Lie algebra automorphism.

Remark 4.8 For every $a \in G$, we write

$$(ada)_* \equiv Ada, \quad i.e.\ a \mapsto Ada \tag{4.32}$$

and is called the **Adjoint Representation of** G.

Problem 4.22 *Is Ada invertible?*

Solution : *By virtue of the last theorem, we find*

$$Ada \equiv (ada)_* : \mathbf{g} \to \mathbf{g}$$

is a Lie algebra homomorphism. Further $\mathbf{g} \cong T_e(G)$, where $T_e(G)$ is a finite-dimensional vector space. Thus, Ada is a linear transformation from $T_e(G) \to T_e(G)$. Again, for every $a \in G$, we obtain

$$
\begin{aligned}
(Ada) \circ (Ada^{-1}) &= (ada)_* \circ (ada^{-1})_* \\
&= (L_a \circ R_{a^{-1}}) \circ (R_a \circ L_{a^{-1}})_*, \ \ by \ (4.29) \\
&= Ade
\end{aligned}
$$

Similarly, it can be shown that $(Ada^{-1}) \circ (Ada) = Ade$. Hence, $\mathbf{Ada}^{-1} = (\mathbf{Ada})^{-1}$. This completes the proof.

4.5 One-Parameter Subgroup of a Lie Group

Let G be a Lie group and a mapping $a : \mathbb{R} \to G$ denoted by $a : t \mapsto a(t)$ be a differentiable curve on G. If for all t, s in $\mathbb{R}$

$$a(t + s) = a(t)a(s), \tag{4.33}$$

then the family $\{a(t) | t \in \mathbb{R}\}$ is called a **one-parameter subgroup** of G.
Exercises

Exercise 4.19 *Let $H = \{a(t) | t \in \mathbb{R}\}$ be a one-parameter subgroup of a Lie group G. Show that H is a commutative subgroup of G.*

Exercise 4.20 *If X is a left invariant vector field on a Lie Group G, prove that X is complete.*

Theorem 4.6 *Let X be the generator generated by one-parameter group of transformations R_{a_t} and let Y be that of L_{a_t}. Then X is left invariant and Y is right invariant and $X_e = Y_e = a'(0)$ hold, where $a'(t)$ denotes the tangent vector to the curve $\boldsymbol{a}$ at $a(t)$.*

Proof For $h \in G$, let $L_h(p) = q$. Therefore, $p = h^{-1}q$. Now from (2.34)

$$\{(L_h)_* X_p\} f = X_p(f \circ L_h), \quad \forall f \in F(G)$$
$$= X_{h^{-1}q}(f \circ L_h)$$
$$= \lim_{t \to 0} \frac{1}{t}[(f \circ L_h)\{R_{a_t}(h^{-1}q)\} - (f \circ L_h)(h^{-1}q)], \text{ by } (2.47)$$
$$= \lim_{t \to 0} \frac{1}{t}\{(f \circ R_{a_t})(q) - f(q)\}, \text{ by } (4.2), (4.4)$$
$$= X_q f, \text{ by } (2.47)$$
$$\therefore \quad (L_h)_* X_p = X_q, \quad \forall f$$
$$\text{or } (L_h)_* X_p = X_{L_h(p)},$$

which shows that X is a left invariant vector field on Lie group G (refer to (4.8)). Similarly, for $h \in G$, let $R_h(p) = q$, therefore, $p = h^{-1}q$. In view of (2.34), we have

$$\{(R_h)_* Y_p\} f = Y_p(f \circ R_h) \forall f \in F(G)$$
$$= Y_{h^{-1}q}(f \circ R_h)$$
$$= \lim_{t \to 0} \frac{1}{t}[(f \circ R_h)\{L_{a_t}(qh^{-1})\} - (f \circ R_h)(qh^{-1})]$$
$$= \lim_{t \to 0} \frac{1}{t}\{(f \circ L_{a_t})(q) - f(q)\}, \text{ by } (4.3), (4.4)$$
$$= Y_q f$$
$$\therefore \quad (R_h)_* Y_p = Y_q, \quad \forall f$$
$$= Y_{R_h(p)}.$$

Thus, by (4.9), we can say that Y is right invariant. Since $R_{a_t}(e) = a_t = a(t)$, $a(t)$ is an integral curve of X and hence $X_{a(t)} = a'(t)$ holds. In particular, $X_e = a'(0)$. By similar manner, $Y_e = a'(0)$.

Theorem 4.7 *Let $\{\phi_t | t \in \mathbb{R}\}$ be a one-parameter group of transformations on a Lie group G generated by a left invariant vector field and*

$$\phi_t(e) = a(t) = a_t.$$

If for every $s \in G$, $\phi_t \circ L_s = L_s \circ \phi_t$, for all $s \in G$, then the set $\{a(t) | t \in \mathbb{R}\}$ is a one-parameter subgroup of G and $\phi_t = R_{a_t}$ holds for all $t \in \mathbb{R}$. If $\phi_t \circ R_s = R_s \circ \phi_t$ holds for all $s \in G$, then the set $\{a(t) | t \in \mathbb{R}\}$ is a one-parameter subgroup of G and $\phi_t = L_{a_t}$ holds for all $t \in \mathbb{R}$

Proof As defined

$$a(s+t) = \phi_{s+t}(e) = \phi_{t+s}(e), \ \forall s, t \in \mathbb{R}$$
$$= \phi_t(\phi_s(e)), \ \text{as } \{\phi_t | t \in \mathbb{R}\} \text{ is one-parameter group of transformations}$$
$$= \phi_t a(s)$$
$$= \phi_t(L_{a(s)} e)$$
$$= (L_{a(s)} \circ \phi_t)(e), \ \text{from the hypothesis, as } a(s) \in G$$
$$= L_{a(s)}(a(t))$$
$$= a(s)a(t).$$

So from (4.33), $\{a(t) | t \in \mathbb{R}\}$ is a one-parameter subgroup of G. Also

$$\phi_t(s) = \phi_t(se) = (\phi_t \circ L_s)(e) = (L_s \circ \phi_t)(e) = L_s(a_t) = sa_t = R_{a_t}(s).$$

Therefore, $\phi_t = R_{a_t}$. Similarly, it can be shown that $\phi_t = L_{a_t}$ when $\phi_t \circ R_s = R_s \circ \phi_t$ holds. This completes the proof.

For each $X \in \mathbf{g}$, we set

$$\phi_t(e) = R_{a_t}(e) = a_t = a(t) = \exp(tX), \tag{4.34}$$

where $\{\phi_t | t \in \mathbb{R}\}$ is the one-parameter group of transformations on G generated by X. We call $\{a_t = a(t) = \phi_t(e)\}$ the **one-parameter subgroup** of G, generated by X.

The map $X \to \exp X$, is a map from $\mathbf{g}$ to G and is said to be the **exponential map**.

Problem 4.23 *Let G be a Lie group. For every $X \in \mathbf{g}$, let Y be the generator induced by the one-parameter group of transformations $\{\phi_t | t \in \mathbb{R}\}$ defined by*

$$\phi : G \to G, \phi(t, x) \equiv \phi_t(x) = \exp(tX)x, \ \forall x \in G.$$

Prove that Y is right invariant.

Solution : *We have to prove that, for every $a \in G$, $(R_a)_* Y = Y$. In view of Exercise 2.48, we are to prove that $(R_a \circ \phi_t) = (\phi_t \circ R_a)$. Now*

$$(R_a \circ \phi_t)(x) = R_a(\exp(tX)x), \quad \text{by hypothesis}$$
$$= \exp(tX)R_a x$$
$$= \exp(tX)xa, \quad \text{by (4.3).}$$

Also, $(\phi_t \circ R_a)(x) = \phi_t(xa) = \exp(tX)xa$. Thus, $(R_a \circ \phi_t) = (\phi_t \circ R_a)$ i.e. $(R_a)_* Y = Y$. Hence, the vector field Y is right invariant.

Problem 4.24 *Let G be an Abelian Lie group. Prove that $[X, Y] = 0$, where X, Y are left invariant vector fields on G.*

Solution : *From Problem 4.23, for* $X, Y \in g$, *let*

$$\phi_t(x) = \exp(tX)x,$$
$$\psi_s(x) = \exp(sY)x, \ t, s \in \mathbb{R}.$$

By virtue of Theorem 2.8, we know that if $\{\phi_t | t \in \mathbb{R}\}$ *is the one-parameter group of transformations generated by* X, *then for every vector field* Y,

$$[X, Y] = \lim_{t \to 0} \frac{1}{t} \{Y - (\phi_t)_* Y\}$$

$$or \ \ [X, Y]_q = \lim_{t \to 0} \frac{1}{t} \{Y_q - ((\phi_t)_* Y)_q\}, \ \ q \in G.$$

Let us write $\phi_t(p) = q$, $p \in G$, *then* $p = \phi_{-t}(q)$ *and hence*

$$((\phi_t)_* Y)_q = ((\phi_t)_* Y)_{\phi_t(p)} = (\phi_t)_* Y_p [refer to (2.35)] = (\phi_t)_* Y_{\phi_{-t}(q)}.$$

Thus, $[X, Y] = 0$ *if* $((\phi_t)_* Y)_q = Y_q$ *i.e.* $(\phi_t)_* Y_{\phi_{-t}(q)} = Y_q$, *i.e. we have to show that* Y *is invariant under* ϕ_t. *Hence, from Exercise 2.48, we wish to prove* $\phi_t \circ \psi_s = \psi_s \circ \phi_t$. *Now*

$$(\phi_t \circ \psi_s)(x) = \phi_t(\exp(sY)x)$$
$$= \exp(sY)\exp(tX)x$$
$$= \exp(tX)\exp(sY)x, \ \ as \ G \ is \ abelian$$
$$= \psi_s(\exp(tX)x)$$
$$= (\psi_s \circ \phi_t)(x).$$

Therefore, $\phi_t \circ \psi_s = \psi_s \circ \phi_t$, *for all* $x \in G$. *This completes the proof.*

Theorem 4.8 *If* $X, Y \in g$, *then* $[Y, X] = \lim_{t \to 0} \frac{1}{t} \{(Ada_t^{-1})Y - Y\}$.

Proof If $\{\phi_t | t \in \mathbb{R}\}$ is the one-parameter group of transformations on a Lie group generated by the left invariant vector field X, then

$$[Y, X] = \lim_{t \to 0} \frac{1}{t} \{(\phi_t)_* Y - Y\}, \ \ \text{for every vector field } Y \in \chi(G).$$

Now $Ada_t^{-1} = (ada_t^{-1})_* = (R_{a_t} \circ L_{a_t^{-1}})_*$, see (4.29).
    If $Y \in g$, then from above after a few steps,

$$(Ada_t^{-1})Y = (R_{a_t})_* Y = (\phi_t)_* Y, \ \ \text{from Theorem 4.7.}$$

Thus

$$[Y, X] = \lim_{t \to 0} \frac{1}{t}\{(Ada_t^{-1})Y - Y\}.$$

This completes the proof.

4.6   Lie Transformation Group (Action of a Lie Group on a Manifold)

A Lie Group G is a Lie Transformation Group on a manifold M or G is said to act differentiably on M if the following conditions hold:

$$\begin{cases} (i)\ (a, p) : G \times M \to pa(\in M) \text{ is a differentiable map;} \\ (ii)\ \text{Each } a \in G \text{ induces a transformation on } M, \text{ denoted by } p \mapsto pa \quad (4.35) \\ (iii)\ p(ab) = (pa)b,\ \forall a, b \in G. \end{cases}$$

We say that G **acts on** M **on the right** as (i) and (iii) can be written as

(i) $R_a p = pa.$
(ii) $R_{ab} p = p(ab) = (pa)b.$

Similarly, the action of G on M on the left can be defined.
Exercises

Exercise 4.21 *Let* $G = GL_2(\mathbb{R})$, $M = \mathbb{R}$ *and* $\theta : G \times M \to M$ *be a differentiable mapping defined by*

$$\theta\left(\begin{pmatrix} a & b \\ 0 & 1 \end{pmatrix}, p\right) = ap + b, \ a > 0, a, b \in \mathbb{R}.$$

Show that θ *is an action on* M.

If G acts on M on the right such that $pa = p$, $\forall p \in M$ implies that $a = e$, then G is said to act **effectively** on M.

However, if G acts on M on the right such that $pa = p$, $\forall p \in M$ implies that $a = e$, for some $p \in M$, then G is said to act **freely** on M.

Theorem 4.9 *If* G *acts on* M, *then the mapping* $\sigma : g \to \chi(M)$ *denoted by* $\sigma : A \to \sigma(A) = A^*$ *is a Lie algebra homomorphism. It is to be noted that* $\sigma(A) = A^*$ *is called the* **fundamental vector field on** M, *corresponding to* $A \in g$ *(Fig. 4.2).*

Proof For every $p \in M$, let $\sigma_p : G \to M$ be a mapping such that

$$\sigma_p(a) = pa, \ \forall a \in G. \tag{4.36}$$

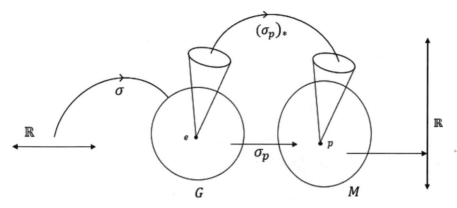

Fig. 4.2 Lie Algebra Homomorphism

Note that both **g**, $\chi(M)$ are algebra, hence we have to show

(i) σ is linear.
(ii) $\sigma([A, B]) = [\sigma(A), \sigma(B)]$.

Note that every $A \in \mathbf{g}$ induces $\{\phi_t(e) \mid t \in \mathbb{R}\}$ as its one-parameter group of transformations on G such that

$$a(t) = a_t = \phi_t(e).$$

The map $(\sigma_p)_* : T_e(G) \to T_{\sigma_p(e)}(M) = T_p(M)$ is a linear map.
    Now by hypothesis, we have

$$(\sigma_p)_* A_e = \{\sigma(A)\}_{\sigma_p(e)} = \{\sigma(A)\}_p = A_p^*. \tag{4.37}$$

For every $A, B \in \mathbf{g}$, we have $A + B \in \mathbf{g}$ and

$$\{\sigma(A + B)\}_p = (\sigma_p)_*(A + B)_e = (\sigma_p)_*(A_e + B_e) = \{\sigma(A)\}_p + \{\sigma(B)\}_p$$
$$\{\sigma(bA)\}_p = (\sigma_p)_* A_e = b\{\sigma(A)\}_p, \ \forall b \in \mathbb{R}.$$

This proves that σ is linear.
    Again, A_e is the tangent vector to the curve $a(t) \equiv a_t$ at $a(0) = e$. Then $(\sigma_p)_* A_e$ is the tangent vector to the curve

$$\sigma_p(a_t) = pa_t, \ \text{by (4.36)}$$
$$= R_{a_t}(p), \ \text{by (4.3)}$$

at $\sigma_p(a_0) = \sigma_p(e) = p$. Thus, A_p^* induces $\{R_{a_t}(p)\}$ as its one-parameter group of transformations on M. Now

$$[\sigma(A), \sigma(B)]_p = [A^*, B^*]_p$$
$$= \lim_{t \to 0} \frac{1}{t} \{ B_p^* - ((R_{a_t})_* B^*)_p \}, \text{ by Theorem 2.8}$$
$$= \lim_{t \to 0} \frac{1}{t} \{ (\sigma_p)_* B_e - (R_{a_t})_* B_q^* \}, \text{ say, where } R_{a_t}(q) = p,$$

$i.e.\ q = pa_t^{-1}.$

$$\therefore (R_{a_t})_* B_q^* = (R_{a_t})_* B_{pa_t^{-1}}^* = (R_{a_t} \circ \sigma_{pa_t^{-1}})_* B_e, \text{ by (4.37).}$$

Now, $R_{a_t} \circ \sigma_{pa_t^{-1}} : G \to M$ and hence for $b \in G$, we have

$$(R_{a_t} \circ \sigma_{pa_t^{-1}})(b) = pa_t^{-1} ba_t = \sigma_p((ada_t^{-1})(b)) = (\sigma_p \circ ada_t^{-1})(b).$$
$$\therefore R_{a_t} \circ \sigma_{pa_t^{-1}} = \sigma_p \circ ada_t^{-1}.$$

Hence

$$(R_{a_t})_* B_q^* = (\sigma_p)_* ((ada_t^{-1})_* B_e) = (\sigma_p)_* ((Ada_t^{-1}) B_e), \text{ by (4.32).}$$

Thus

$$[\sigma(A), \sigma(B)]_p = \lim_{t \to 0} \frac{1}{t} \{ (\sigma_p)_* B_e - (\sigma_p)_* ((Ada_t^{-1}) B_e) \}$$
$$= (\sigma_p)_* [A, B]_e, \text{ as } (\sigma_p)_* \text{ is linear and by Theorem 4.8}$$
$$= (\sigma([A, B]))_p, \text{ by (4.37)}$$

$i.e.\ \sigma([A, B]) = [\sigma(A), \sigma(B)].$

Hence, the mapping $\sigma : \mathbf{g} \to \chi(M)$ is a linear algebra homomorphism. This completes the proof.

Theorem 4.10 *If G acts effectively on M, then the map* $\sigma : \mathbf{g} \to \chi(M)$ *defined by* $A \mapsto \sigma(A) = A^*$ *is an isomorphism.*

Proof In view of Theorem 4.9, we can say that σ is a Lie algebra homomorphism. We are left to prove that σ is bijective. Let $\sigma(A) = \sigma(B)$ hold for every $A, B \in \mathbf{g}$. Then,

$$\sigma(A - B) = \theta \Rightarrow (A - B)^* = \theta.$$

Now for every A, B in $\mathbf{g}$, $A - B(\in \mathbf{g})$ will generate $\{\psi_t(e) | t \in \mathbb{R}\}$(say) as its one-parameter group of transformations on G such that $(A - B)_e$ is the tangent vector to the curve, given by

$$b_t = \psi_t(e) \text{ at } \psi_0(e) = b_0 = e. \tag{4.38}$$

Consequently, $(\sigma_p)_*(A - B)_e$ is the tangent vector to the curve

$$\sigma_p(b_t) = pb_t = R_{b_t}(p), \text{ at } p b_0 = p.$$

Thus, $(A - B)^* \equiv (\sigma_p)_*(A - B)_e$ will generate $\{R_{b_t}(p)|\, t \in \mathbb{R}\}$ as its one-parameter group of transformations on M. But the integral curve of the null-vector $(A - B)^*$ will reduce to the initial point itself, refer to Problem 2.42. Hence,

$$\sigma_p(b_t) = p \quad i.e. \quad R_{b_t}(p) = p \quad i.e. \quad p\,b_t = p.$$

As G acts effectively on M, from the foregoing equation, we obtain $b_t = e$, $\forall\, p \in M$. Again, $(L_p)_*(A - B) = A - B$. Thus, using Exercise 2.48, we have $L_p \circ \psi_t = \psi_t \circ L_p$. Hence

$$\begin{aligned}
\psi_t(q) = \psi_t(qe) = (\psi_t \circ L_q)(e) &= L_q(\psi_t(e)) \\
&= L_q(b_t), \ \text{refer to (4.38)} \\
&= q, \ \text{from above.}
\end{aligned}$$

From the definition, we have

$$(A - B)_q f = \lim_{t \to 0} \frac{1}{t}\{f(\psi_t(q)) - f(q)\} = 0.$$

Therefore, $A - B$ is a null vector. So from $\sigma(A) = \sigma(B)$, we must have $A = B$. Clearly, σ is surjective. Hence, σ is injective. Thus, σ is bijective and consequently, σ is an isomorphism.

Exercises

Exercise 4.22 *If G acts freely on M, the proof that for every non-null left invariant vector field A, the fundamental vector field A^* can never vanish.*

Exercise 4.23 *Prove that the map $\theta : \mathbb{R}^+ \times \mathbb{R} \to \mathbb{R}$ defined by $(a, p) \mapsto ax$ is an action of $\mathbb{R}^+$ on $\mathbb{R}$. Is it free?*

Answer/Hint

4.22. Use Theorems 4.9 and 4.10.
4.23. No

References

1. Abraham, R., Marsden, J.E., Ratiu, T.: Manifolds, Tensor Analysis and Applications, 2nd edn. Springer (1983)
2. Bishop, R.L., Crittenden, R.J.: Geometry of Manifolds. AMS Chelsea Publishing, Providence (2001)
3. Boothby, W.M.: An Introduction to Differentiable Manifolds and Riemannian Geometry, 2nd edn. Academic Press, New York (2002)
4. Castillo, G.F.: Differentiable Manifolds. Birkhäuser (2012)
5. Curtis, W.D., Miller, F.R.: Differential Manifolds and Theoretical Physics. Academic Press (1985)
6. Fitzpatrick, P.M.: Advanced Calculus. Pure and Applied, Undergraduate Texts, 2nd edn. American Mathematical Society, Providence, Rhode Island
7. Gadea, P.M., Masqué, J.M., Mykytyuk, I.V.: Analysis and Algebra on Differentiable Manifolds, 2nd edn. Springer (2013)
8. Gallot, S., Hulin, D., Lafontaine, J.: Riemannian Geometry. Springer, Berlin (2004)
9. Guillemin, V., Pollack, A.: Differential Topology. Prentice-Hall, New Jersey (1974)
10. Helgason, S.: Differential Geometry, Lie Groups and Symmetric Spaces. Graduate Studies in Mathematics, vol. 34. Amc. Math. Soc, Providence (2012)
11. Hicks, N.J.: Notes on Differential Geometry. Von Nostrand Reinhold, London (1965)
12. Hoffman, K., Kunze, R.: Linear Algebra, 2nd edn. Pearson (2015)
13. Kobayashi, S., Nomizu, K.: Foundations of Differential Geometry, vol. I, II. Wiley, New York (1996)
14. Kumaresan, S.: A Course in Differential Geometry and Lie Groups. Hindustan Book Agency (2002)
15. Lee, J.M.: Introduction to Smooth Manifolds, 2nd edn. Springer, New York (2012)
16. Matsushima, Y.: Differentiable Manifold. Mercel Dekker INC, New York (1972)
17. Millman, R.S., Parker, G.D.: Elements of Differential Geometry. Pearson (1977)
18. Morita, S.: Geometry of Differential Forms. A.M.S. (2001)
19. Mukherjee, A.: Differential Topology. Hindustan Book Agency (2015)
20. Munkres, J.: Analysis on Manifolds. Addison-Wesley, CA (1991)
21. Munkres, J.: Topology, 2nd edn. Pearson (2000)

© The Editor(s) (if applicable) and The Author(s), under exclusive license to Springer 209
Nature Singapore Pte Ltd. 2023
M. Majumdar and A. Bhattacharyya, *An Introduction to Smooth Manifolds*,
https://doi.org/10.1007/978-981-99-0565-2

22. Prakash, N.: Differential Geometry, An Integrated Approach. Tata-McGraw-Hill Pub. Com Ltd, New-Delhi (1981)
23. Pressley, A.: Elementary Differential Geometry, 2nd edn. Springer (2010)
24. Rudin, W.: Principles of Mathematical Analysis, 3rd edn. Mc.Graw-Hill, Inc
25. Schutz, B.: Geometrical Methods of Mathematical Physics. Cambridge Univ. Press (1980)
26. Sen, S.: A Short Course on Differentiable Manifolds. University of Calcutta (2011)
27. Shirali, S., Vasudeva, H.L.: Multivariable Analysis. Springer (2011)
28. Sinha, B.B.: An Introduction to Modern Differential Geometry. Kalyani Publisher, New-Delhi (1982)
29. Spivak, M.: Calculus on Manifolds. Benjamin, New York (1965)
30. Spivak, M.: Differential Geometry, vol. 1–5, 3rd edn. Publish or Perish Wilmington (1999)
31. Sinha, R.: Smooth Manifolds. Springer (2014)
32. Tu, L.W.: An Introduction to Smooth Manifolds, 2nd edn. Springer, Berlin (2008)
33. Warner, F.W.: Foundations of Differentiable Manifolds and Lie Groups. Springer, Berlin (2010)
34. Weintraub, S.: Differential Forms, Theory and Practice, 2nd edn. Academic Press (2014)
35. Yano, K., Kon, M.: Structures on Manifolds. Worlds Scientific (1984)

Printed in the United States
by Baker & Taylor Publisher Services